T0201258

# QUANTUM DYNAMICS
# FOR CLASSICAL
# SYSTEMS

# QUANTUM DYNAMICS FOR CLASSICAL SYSTEMS

## With Applications of the Number Operator

**FABIO BAGARELLO**

DIEETCAM
University of Palermo
Palermo, Italy

A JOHN WILEY & SONS, INC., PUBLICATION

Cover Design: Michael Rutkowski
Cover Image: © Dmitrii Brighidov/iStockphoto

Published by John Wiley & Sons, Inc., Hoboken, New Jersey
Published simultaneously in Canada

For general information on our other products and services or for technical support, please contact
our Customer Care Department within the United States at (800) 762-2974, outside the United
States at (317) 572-3993 or fax (317) 572-4002.

Wiley also publishes its books in a variety of electronic formats. Some content that appears in print
may not be available in electronic formats. For more information about Wiley products, visit our
web site at www.wiley.com.

*Library of Congress Cataloging-in-Publication Data:*

Bagarello, Fabio, 1964- author.
  Quantum dynamics for classical systems : with applications of the number
operator / Fabio Bagarello.
      pages cm
    Includes bibliographical references and index.
    ISBN 978-1-118-37068-1 (hardback)
  1. Social sciences–Mathematics. 2. Business mathematics. 3. Quantum
theory–Mathematics. I. Title.
  H61.25.B34 2013
  300.1'53012–dc23

                                                          2012020074

Printed in the United States of America

ISBN: 9781118370681

10 9 8 7 6 5 4 3 2 1

Science never solves a problem
without creating ten more.

**George Bernard Shaw**

The most exciting phrase to hear in science,
the one who heralds new discoveries,
is not *Eureka* but *That's funny*...

**Isaac Asimov**

# CONTENTS

# PREFACE

In 2005 or so, I started wondering whether that particular part of quantum mechanics that theoretical physicists call *second quantization* could be used in the analysis of some particular, somehow *discrete*, classical system. In particular, I started considering stock markets, or a reasonably simplified version of these, since at that time this was a very fashionable topic: econophysics was in its early years, and the general feeling was that there was still room, and need, for many other contributions from physicists and mathematicians. I got the idea that the analysis of the huge amount of information going around a real market was only part of what was interesting to do. I was much more interested in considering the viewpoint of the single trader, who is more likely interested in having some control of his own portfolio. Therefore, I constructed a model of a simplified market, just to see if this *strange* approach could be interesting for such a hypothetical trader, and I suddenly realized that "yes, it might make sense to carry on in this line of research, but, wow, it is hard to have such a paper accepted in a good journal." However, after a few weeks, I also realized that this topic seemed to be interesting not only for me, but also for a large community of scientists, and that this community was increasing very fast, producing more and more contributions on the *ArXiv*. People started citing my first paper, and I was contacted by people interested in what I was doing and who wanted to discuss my point of view. This pushed me in the direction of considering more sophisticated models for stock markets, using my knowledge of quantum mechanics for systems with infinite degrees of freedom in this other, and apparently completely different, field. I thought that this was essentially the end of the story: quantum versus economics. Unexpectedly, a few years ago during a conference in Acireale where I gave a talk on my *quantum stock markets*, I had a discussion with an old friend of mine, Franco Oliveri, and he suggested using the same general strategy in a different classical context. I remember that in our first discussion, we were thinking of foxes and chicken, a topic that was not very exciting for me. After a while, we realized that what we were discussing could also have been used to describe something completely different: a love story. And that was the beginning of a long story that still continues. Since then, we have constructed several models for different classical systems, playing with our understanding of these systems and looking for some phenomenological description. It turned out that these models quite often produce nontrivial and, in my opinion, quite interesting features that are not fully explored yet. Moreover, what is also very intriguing to me is that the same general framework can be used in many different contexts, giving rise to a sort of unifying setting.

This book might be considered as a first attempt to summarize what I have done so far in this field. My idea was to make the book *reasonably simple* and self-contained. This is because I expect that some not necessarily mathematical-minded readers might be intrigued by the title, and I do not want to lose these readers. However, a complex system can be made easy only up to a certain extent, and this is exactly what I have tried to do in these notes. Even the love story I will consider in Chapter 3, which from the purely dynamic point of view is surely the simplest situation, is not simple at all. This is not a big surprise, as almost every lover knows very well from personal experience. I should also clarify that it is not my main interest to discuss the *psychological aspects* behind a love story, a migration process, or the choices of traders in a stock market. I am not even interested in giving any abstract, or too general, description of these systems. Here I want to be quantitative! I want to deduce formulas from general ideas, and I want to see what these formulas imply for the system I have in mind, and if they have some predictive power. However, this ultimate goal implies some effort to the reader, who is required to create his own background on quantum mechanics (if needed) by reading Chapter 2. Dear reader, if you can understand Chapter 2, you can understand the rest! On the other hand, if Chapter 2 is too technical for you, do not worry: you could still try to read the book, simply jumping over this chapter. Of course, if you are not a physicist, you will lose a lot. But you can still get the feeling of what is going on. It is up to you! I really hope you enjoy reading this book!

FABIO BAGARELLO

# ACKNOWLEDGMENTS

It is always a pleasure to thank old friends of mine such as Camillo Trapani and Franco Oliveri for their very precious help in so many different situations that I cannot even remember myself. This could be seen as (the first) evidence of the death of my neurons. But, do not worry! I have still enough neurons left in my brain to remember that if something in functional analysis is not particularly clear, Camillo is the right one! And I also have a post-it on my desk that says "Any numerical problem? Call Franco!" What is more funny is that they still answer my questions, even when they are very, very busy! Franco is also responsible, in part, for what is discussed in these notes, and I also thank him for this scientific collaboration and for his enthusiasm.

It is also a pleasure to thank the various editors and referees whom I have *met* along these years, including the ones who referred this book (before it became a book!). Most of them have contributed significantly to the growth of my research, with many useful, and sometimes unexpected, suggestions. Particular thanks goes to Wiley, for the enthusiasm shown for my manuscript.

I dedicate this book to my beloved parents Giovanna and Benedetto, to my brother Vincenzo, and, *dulcis in fundo,* to Federico, Giovanna, and Grazyna. When I look at them I often ask myself the same question, a question for which I have no answer, yet: how can they resist with so much mathematics and physics going around?

# CHAPTER 1

# WHY A QUANTUM TOOL
# IN CLASSICAL CONTEXTS?

Of course, there is no other way to begin this book. In our experience, this is the first question that a referee usually raises when he receives a paper of ours. Hence this is the question that we try to answer in this chapter, to motivate our analysis.

Taking a look at the index, we see that the applications discussed here cover a large variety of problems, from love affairs to migration, from competition between species to stock markets. First of all, we have to stress that we are not claiming that, for instance, a love affair has a quantum nature! (Even though, as every lover knows very well, each love story is surely characterized by a strong stochastic ingredient that could be analyzed, for instance, using tools from probability theory. It is not surprising, then, that one could try to use quantum mechanics as well, in view of its probabilistic interpretation.) Therefore, we are not going to discuss any *quantum love affair*. Rather, we just claim that some quantum tools, and the number operator in particular, can be used, *cum grano salis*, in the analysis of several dynamical systems in which the variables are seen as operator-valued functions. The interesting fact is that the results we deduce using these tools describe very well the dynamics of the system we are considering. This is shown to be true for love affairs first, but this

*Quantum Dynamics for Classical Systems: With Applications of the Number Operator*,
First Edition. Fabio Bagarello.
© 2013 John Wiley & Sons, Inc. Published 2013 by John Wiley & Sons, Inc.

same conclusion apparently holds in other, completely different, contexts (migrations, stock markets, competition between species, etc.).

However, to answer in more detail the question raised in the title, we need a long introduction, and this is the main content of this chapter. We begin with a few useful facts on (anti-)commutation rules, which are used to motivate our answer. Then, we describe briefly other appearances of quantum mechanics in the description of classical systems, proposed by several authors in recent years. We conclude the chapter with the plan of the book.

## 1.1   A FIRST VIEW OF (ANTI-)COMMUTATION RULES

In many fields of quantum mechanics of systems with few or many degrees of freedom, the use of annihilation or creation operators, and of their related number operators, has been proved to be very useful. The first explicit application of the so-called canonical commutation relations (CCR) is usually found at the first level degree in physics while studying the one-dimensional quantum harmonic oscillator (Merzbacher, 1970). CCR are used to produce a purely algebraic procedure that helps in finding the eigenvalues and the eigenvectors of the energy operator $H = \frac{1}{2}\left(p^2 + x^2\right)$ of the oscillator, expressed here in convenient units. This procedure is much simpler than the one that returns the explicit solution of the Schrödinger equation in configuration space (i.e., in terms of the position variable $x$). The quantum nature of the system is reflected by the fact that $p$ and $x$ do not commute. Indeed, they satisfy the following rule: $[x, p] := xp - px = i\,\mathbb{1}$, $\mathbb{1}$ being the identity operator in the Hilbert space $\mathcal{H} = \mathcal{L}^2(\mathbb{R})$ where the oscillator lives. This means that $x$ and $p$ are not classical functions of time but, rather, operators acting on $\mathcal{H}$. As life is usually not easy, and science is even harder, $x$ and $p$ are unbounded operators. This produces a number of extra difficulties, mainly on the mathematical side, which we try to avoid as much as possible in these notes, but which is necessary at least to mention and to have in mind.[1] Introducing $a := \frac{1}{\sqrt{2}}(x + ip)$, its adjoint $a^\dagger := \frac{1}{\sqrt{2}}(x - ip)$, and $N := a^\dagger a$, we can rewrite the Hamiltonian as $H = N + \frac{1}{2}\,\mathbb{1}$, and, if $\varphi_0$ is a vector of $\mathcal{H}$, which is annihilated by $a$, $a\varphi_0 = 0$, then, calling $\varphi_n := \frac{1}{\sqrt{n!}}(a^\dagger)^n \varphi_0$, $n = 0, 1, 2, \ldots$, we have $H\varphi_n = (n + 1/2)\varphi_n$. In the literature, $\varphi_0$ is called

---

[1] Along this book, we add some remarks concerning the unboundedness of some operators used in the description of the system under investigation.

*the vacuum* or *the ground state* of the harmonic oscillator. Then we find, avoiding the hard explicit solution of the Schrödinger differential equation

$$i\frac{\partial \Psi(x,t)}{\partial t} = \frac{1}{2}\left(-\frac{\partial^2 \Psi(x,t)}{\partial x^2} + x^2 \Psi(x,t)\right),$$

the set of eigenvalues ($E_n = n + 1/2$) and eigenvectors ($\varphi_n$) of $H$. In the derivation of these results, the crucial ingredient is the following commutation rule: $[a, a^\dagger] = \mathbb{1}$, easily deduced from $[x, p] = i\mathbb{1}$ and from the definitions of $a$ and $a^\dagger$, which have the useful consequence $[N, a^{\dagger^n}] = na^{\dagger^n}$, $n = 0, 1, 2, \ldots$. Using this result, $H\varphi_n = E_n\varphi_n$ follows immediately.

Standard quantum mechanical literature states a simple extension of $[a, a^\dagger] = \mathbb{1}$ is found soon after the one-dimensional harmonic oscillator, while moving to higher dimensional systems. In this case, the CCR look like $[a_l, a_n^\dagger] = \mathbb{1}\,\delta_{l,n}$, $l, n = 1, 2, 3, \ldots L$: we have $L$ independent modes. It might be interesting to remind that $L$th dimensional oscillators are usually the key ingredients to set up, both at a classical and at a quantum level, many perturbation schemes that are quite useful whenever the dynamics of the system cannot be deduced exactly. Some quantum perturbation approaches are quickly reviewed in Chapter 2 and used all along the book.

Studying quantum field theory, one is usually forced to consider mainly two different kinds of particles, which obey very different commutation rules and, as a consequence of the spin-statistic theorem, different statistics: the *bosons* and the *fermions*. Bosons are particles with integer spin, such as the photons. Fermions are particles with half-integer spin, such as the electrons. This difference in the value of the spin has an important consequence: fermions satisfy the Pauli exclusion principle, whereas bosons do not. This is reflected, first of all, by the wave function that describes any set of identical fermions, which has to be antisymmetric with respect to the change of their variables, or by the wave function for the bosons, which has to be symmetric. Hence, if two indistinguishable fermions have exactly the same quantum numbers (e.g., they occupy the same position in space and they have the same energy), their wave function collapses to 0: such a configuration cannot occur! This is the Pauli exclusion principle, which, of course, does not hold for the bosons. In *second quantization*, (Roman, 1965), the bosons are created by the operators $a_l^\dagger$ and annihilated by their conjugate, $a_l$. Together, they satisfy the CCR above. Analogously, fermions are annihilated and created by similar operators, $b_k$ and $b_k^\dagger$, but these satisfy a different rule, the so-called anticommutation relation (CAR): $\{b_l, b_k^\dagger\} = b_l b_k^\dagger + b_k^\dagger b_l = \mathbb{1}\,\delta_{l,k}$, with $\{b_l, b_k\} = \{b_l^\dagger, b_k^\dagger\} = 0$, $l, k = 1, 2, 3, \ldots$. The main difference between these two commutation rules is easily understood. While the operator $a_l^2$ is different from 0, the square of $b_l$ is automatically 0, together with all its higher powers. This is again an evidence of the Pauli principle: if we try to construct a system with two fermions with the same quantum numbers (labeled by $l$) using the language of second quantization, we should act twice with $b_l^\dagger$ on the vacuum $\varphi_0$, that is, on the vector annihilated by all the $b_l$s. But, as $b_l^{\dagger 2} = 0$, the resulting vector is 0: such a state has probability 0 to occur and, as a consequence, the Pauli principle is preserved!

In Chapter 2, we show, among other things, that the eigenvalues of $N_l^{(a)} := a_l^\dagger a_l$ are $0, 1, 2, \ldots$, whereas those of $N_l^{(b)} := b_l^\dagger b_l$ are simply $0, 1$. This is related to the fact that the *fermionic* and the *bosonic* Hilbert spaces differ as the first one is finite dimensional, whereas the second is infinite dimensional. Needless to say, this produces severe differences from a technical point of view. In particular, operators acting on a (finite modes) fermionic Hilbert space are automatically bounded, whereas those acting on a bosonic Hilbert space are quite often unbounded.

## 1.2 OUR POINT OF VIEW

In many classical problems, the relevant quantities we are interested in change discontinuously. For instance, if you consider a certain population $\mathcal{P}$, and its time evolution, the number of people forming $\mathcal{P}$ cannot change arbitrarily: if, at $t_0 = 0$, $\mathcal{P}$ consists of $N_0$ elements, at $t_1 = t_0 + \Delta t$, $\mathcal{P}$ may only consist of $N_1$ elements, with $N_1 - N_0 \in \mathbb{Z}$. The same happens if our system consists of two (or more) different populations, $\mathcal{P}_1$ and $\mathcal{P}_2$ (e.g., preys and predators or two migrating species): again, the total number of their elements can only take integer values.

Analogously, if we consider what in these notes is called *a simplified stock market* (SSM), that is, a group of people (the *traders*) with some money and a certain number of shares of different kind, which are exchanged between the traders who pay some cash for that, it is again clear that natural numbers play a crucial role: in our SSM, a trader may have only a natural number of shares (30, 5000, or $10^6$, but not 0.75 shares), and a natural number of units of cash (there is nothing $< 1$ cent of euro, for instance). Hence, if two traders buy or sell a share, the number of shares in their portfolios increases or decreases by one unit, and the amount of money they possess also changes by an integer multiple of the unit of cash.

In the first part of these notes, we also consider some quantities that change continuously but that can also still be measured, quite naturally, using discrete values: this is the case, for instance, of the love affair between Alice and Bob described in Chapter 3: in some old papers, see

Strogatz (1988) and Rinaldi (1998a,b) for instance, the mutual affection between the two actors of the love affair is described by means of two continuous functions. However, it is not hard to imagine how a similar description could be given in terms of discrete quantities: this is what we have done, for instance, in Bagarello and Oliveri (2010, 2011): Bob's affection for Alice is measured by a discrete index, $n_B$, which, when it increases during a time interval $[t_i, t_f]$, from, say, a value 1 to the value 2, describes the fact that Bob's love for Alice increases during that particular time interval. Analogously, Alice's affection for Bob can be naturally measured by a second discrete index, $n_A$, which, when its value decreases from, say, 1 to 0, describes the fact that Alice's love for Bob simply disappears.

These are just a few examples, all described in detail in these notes, showing how the use of discrete quantities is natural and can be used in the description of several systems, in very different situations. Of course, at first sight, this may look as a simple discretization of a continuous problem, for which several procedures have been proposed along the years. However, this is not our point of view. We adopt here a rather different philosophy, which can be summarized as follows: the discrete quantities used in the description of the system $S$ under analysis are closely related to the eigenvalues of some self-adjoint operator. Moreover, these operators can be quite often approximated with effective, finite dimensional, self-adjoint matrices, whose dimensions are somehow fixed by the initial conditions; see, for instance, Chapter 3. Then the natural question is the following: how can we deduce the dynamical behavior of $S$? This is, of course, the hard part of the job! Along all our work, we have chosen to use a Heisenberg-like dynamics, or its Schrödinger counterpart, which we believe is a good choice for the following reasons:

1. It is a natural choice when operators are involved. This is exactly the choice used in quantum mechanics, where the Heisenberg representation is adopted in the description of the dynamics of any closed microscopic system.

2. It is quite easy to write down an energy-like operator, the Hamiltonian $H$ of the system $S$, which drives the dynamics of the system. This is due to the fact that, following the same ideas adopted in particle physics, the Hamiltonian contains in itself the phenomena it has to describe. This aspect is clarified first in Sections 3.3 and 3.4 in a concrete situation, while in Chapter 6, we discuss the role and the construction of $H$ in more detail and in a very general condition. Among the other criteria, the explicit definition of $H$ will also be suggested by the existence of some conserved quantities of

$S$: if $X$ is an operator, which is expected to be preserved during the time evolution of $S$, for instance, the total amount of cash in a closed SSM, then, because of the definition of the Heisenberg dynamics, $H$ must commute with $X$: $[H, X] = 0$. This gives some extra hints on how to define $H$ explicitly, and then $H$ can be used to find the time evolution of any observable $A$ of $S$ using the standard prescription: $A(t) = e^{iHt} A(0) e^{-iHt}$, $A(0)$ being the value of $A$ at $t = 0$. We refer to Chapter 2 for many more details on the dynamics of $S$.

3. It produces results which, at least for those systems considered in the first part of these notes, look quite reasonable; that is, they are exactly those results, which one could expect to find as they reproduce what we observe in real life. This is a good indication, or at least gives us some hope, that the dynamics deduced for the systems discussed in Part II of the book, that is, for SSMs, reflect a reasonable time evolution for those systems.

This list shows that we have two technical and one a posteriori reason to use an energy-like operator $H$ to compute the dynamics of $S$. This is not, of course, the end of the story, but, in our opinion, it is already a very good starting point.

## 1.3 DO NOT WORRY ABOUT HEISENBERG!

People with a quantum mechanical background know very well that, whenever incompatible (i.e., not commuting) observables appear in the description of a given physical system $S$, some uncertainty results follow. Hence, one may wonder how our quantum-like description could be compatible with the classical nature of $S$, whose observable quantities are not expected to be affected by any error, except, at most, by the error due to the experimental settings. This problem, actually, does not exist at all in the applications considered in these notes as all the observables we are interested in form a commuting subset of a larger nonabelian algebra. Therefore, they can be diagonalized simultaneously and a common orthonormal (o.n.) basis of the Hilbert space $\mathcal{H}$ used in the description of $S$, made of eigenstates of these observables, can be indeed obtained, as we see several times in Chapters 3–9. This means that, in the complete description of $S$, all the results that are deduced using our approach are not affected by any uncertainty because all the relevant self-adjoint operators whose eigenvalues are relevant to us are compatible, that is, mutually commuting.

It should also be mentioned that, in some specific applications, the impossibility of observing simultaneously two (apparently) classical quantities has been taken as a strong indication of the relevance of a quantum-like structure in the description of that process, showing, in particular, the importance of noncommuting operators. This is what was proposed, for instance, in Segal and Segal (1998), which is based on the natural assumption that a trader in a market cannot know, at the same time, the price of a certain share and its forward time derivative. The reason is clear: if the trader has access to both these information with absolute precision, then he is surely able to earn as much as he wants! For this reason, Segal and Segal proposed to use two noncommuting operators to describe the price and its time derivative. Going back to the title of this section, although in this book we are happy to not deal with the uncertainty principle, in other approaches this is actually seen as the main motivation to use a quantum or noncommutative approach for a macroscopic system. For this reason, also in view of possible future applications, we describe in Section 2.4 a possible mathematical derivation of a rather general inequality for non-commuting operators, which, fixing the operators in a suitable way, gives back the Heisenberg uncertainty relation.

## 1.4  OTHER APPEARANCES OF QUANTUM MECHANICS IN CLASSICAL PROBLEMS

Going back to the crucial aspect of this book, which is surely the mixture of quantum and classical words, we want to stress again that this is surely not the first place in which such a mixture is extensively adopted. On the contrary, in the past few years, a growing interest in classical applications of quantum ideas appeared in the literature, showing that more and more people believe that there is not a really big difference between these two worlds or that, at least, some mathematical tool originally introduced in quantum mechanics may also play a significant role in the analysis of classical systems. These kinds of mixtures can be found in very different fields such as economics (Aerts et al., 2012; Segal and Segal, 1998; Schaden, 2002; Baaquie, 2004; Accardi and Boukas, 2006; Al, 2007; Choustova, 2007; Ishio and Haven, 2009; Khrennikov, 2010; Romero et al., 2011; Pedram, 2012), biology (Engel et al., 2007; Arndt et al., 2009; Pusuluk and Deliduman; Martin-Delgado; Panitchayangkoon et al., 2011; Ritz et al. 2004), sociology, and psychology (Shi, 2005; Jimenez and Moya, 2005; Busemeyer et al., 2006; Khrennikov, 2006; Aerts et al., 2009, 2010; Yukalov and Sornette, 2009a,b; Aerts, 2010; Mensky, 2010; Makowski and Piotrowski, 2011; Yukalov and Sornette), and also in more general contexts (Abbott et al., 2008; Khrennikov, 2010), just to cite a few. The number of scientific contributions having classical applications of quantum mechanics as their main subject is fast increasing. To have an idea of what is going on, it is enough to follow the *arXiv* at xxx.lanl.gov, where almost everyday new papers are submitted. This, of course, provides encouragement to pursue our analysis and to check how far we can go with our techniques and how our results can be used to explain some aspects of the real macroscopic world.

## 1.5   ORGANIZATION OF THE BOOK

This book is essentially organized in three parts. In the first part, Chapter 2, we review some important aspects of quantum mechanics, which are used in the rest of the book. In particular, we discuss the dynamical problem in ordinary quantum mechanics using several representations and describing the relations between them. We also discuss in great detail CCR, CAR, and some perturbative approaches, which are used sometimes in the book, as well as other tools and aspects related to quantum mechanics of certain interest for us, such as the Green's functions, the states over the algebra of bounded operators, and the Heisenberg uncertainty principle.

In the second part, Chapters 3–5, we show how the CCR and the CAR can be used for classical systems with few *degrees of freedom*. In particular, we discuss our point of view on love relations, describing also the role of the environment surrounding the people involved in the love affair. Later, in Chapter 4, we show how the same general framework can be used in the description of competitions between species and for other biological systems. For instance, we describe a migration process involving two populations, one living in a rich area and the second one in a poor region of a two-dimensional lattice.

Chapter 5 is dedicated to the description of the dynamical behavior of a biological-like system (e.g., some kind of bacteria) coupled to two reservoirs, one describing the food needed by the system to survive, and the other mimicking the garbage that is produced by the system itself.

Chapter 6 is a sort of an interlude, useful to fix the ideas on the role of the Hamiltonian of the system $S$ we are describing, and on how this Hamiltonian should be constructed. More explicitly, we identify three main steps in the analysis of $S$: the first step consists in understanding the main mechanisms taking place in $S$, with a particular interest to the interactions between its constituents. Second, we deduce the Hamiltonian for $S$, $H_S$, following a set of rather general rules, which is listed and explained in detail. The final step in our analysis of $S$ is the deduction, from $H_S$, of its dynamics. This is usually the hardest part of the job.

The last part of the book, Chapters 7, 8, and 9, is concerned with systems with many *degrees of freedom* and in particular, with our closed SSM. We propose several models for an SSM, from very simple to more complicated ones, and we consider some of the related dynamical features. In particular, most of the times, we will be interested in the deduction of the time evolution of the portfolio of each single trader, but in Chapter 9, we also compute a *transition probability* between different states of the SSM.

We devote Chapter 10 to some final considerations and to possible generalizations and applications. In particular, we discuss several possible extensions of our main tools, the CCR and the CAR. This could be useful to describe a nonunitary time evolution, describing some decay, as well as systems with a finite, and larger than 2, number of energy levels.

# CHAPTER 2

# SOME PRELIMINARIES

In this chapter, we briefly review some basic facts in quantum mechanics. In particular, we focus on what physicists usually call *second quantization*, which is used in the rest of the book. This chapter is essentially meant to keep these notes self-contained and to fix the notation. Nevertheless, the reader with a background in quantum mechanics is surely in a better position to fully comprehend the material discussed in this book. Of course, people with such a background could safely skip this chapter. It might also be worth stressing that, sometimes, we discuss something more that what is really used, just to give a reasonably closed form to the arguments presented in this chapter.

## 2.1 THE BOSONIC NUMBER OPERATOR

Let $\mathcal{H}$ be a Hilbert space and $B(\mathcal{H})$ the set of all the bounded operators on $\mathcal{H}$. $B(\mathcal{H})$ is a so-called *C\*-algebra*, that is, an algebra with involution that is complete under a norm, $\| \cdot \|$, satisfying the *C\*-property*: $\|A^*A\| = \|A\|^2$, for all $A \in B(\mathcal{H})$. As a matter of fact, $B(\mathcal{H})$ is usually seen as a *concrete realization* of an abstract C\*-algebra. Let $\mathcal{S}$ be our physical system and $\mathcal{A}$ the set of all the operators useful for a complete description of $\mathcal{S}$, which includes the observables of $\mathcal{S}$, that is, those

*Quantum Dynamics for Classical Systems: With Applications of the Number Operator,*
First Edition. Fabio Bagarello.
© 2013 John Wiley & Sons, Inc. Published 2013 by John Wiley & Sons, Inc.

quantities that are measured in a concrete experiment. For simplicity, it would be convenient to assume that $\mathcal{A}$ is a C*-algebra by itself, possibly coinciding with the original set $B(\mathcal{H})$, or, at least, with some closed subset of $B(\mathcal{H})$. However, this is not always possible in our concrete applications. This is because of the crucial role of some unbounded operators within our scheme: unbounded operators do not belong to any C*-algebra. However, if $X$ is such an operator, and if it is self-adjoint, then $e^{iXt}$ is unitary and, therefore, bounded. The norm of $e^{iXt}$ is 1, for all $t \in \mathbb{R}$, and $X$ can be recovered by taking its time derivative in $t = 0$ and multiplying the result by $-i$. For this reason, C*-algebras and their subsets are also relevant for us when unbounded operators appear.

A concrete situation where these kinds of problems arise is the description of the time evolution of $S$, which is driven by a self-adjoint operator $H = H^\dagger$, which is called *the Hamiltonian* of $S$ and which, in standard quantum mechanics, represents the energy of $S$. In most cases, $H$ is unbounded. In the so-called *Heisenberg representation*, see Section 2.3.2, the time evolution of an observable $X \in \mathcal{A}$ is given by

$$X(t) = e^{iHt} X e^{-iHt} \tag{2.1}$$

or, equivalently, by the solution of the differential equation

$$\frac{dX(t)}{dt} = i e^{iHt} [H, X] e^{-iHt} = i [H, X(t)], \tag{2.2}$$

where $[A, B] := AB - BA$ is the *commutator* between $A$ and $B$. In view of what we have discussed before, $e^{\pm iHt}$ are unitary operators, hence they are bounded. Time evolution defined in this way is usually a one parameter group of automorphisms of $\mathcal{A}$:[1] for each $X, Y \in \mathcal{A}$, and for all $t, t_1, t_2 \in \mathbb{R}$, $(XY)(t) = X(t)Y(t)$ and $X(t_1 + t_2) = (X(t_1))(t_2)$. An operator $Z \in \mathcal{A}$ is a *constant of motion* if it commutes with $H$. Indeed, in this case, Equation 2.2 implies that $\dot{Z}(t) = 0$, so that $Z(t) = Z(0)$ for all $t$. It is worth stressing that, in Equations 2.1 and 2.2, we are assuming that $H$ does not depend explicitly on time, which is not always true. We give a rather more detailed analysis of time evolution of a quantum system, under more general assumptions, in Section 2.3.

As already discussed briefly in Chapter 1, a special role in our analysis is played by the CCR: we say that a set of operators $\{a_l, a_l^\dagger, l =$

---

[1]It might happen, for some $H$, that the time evolution of an element of $\mathcal{A}$ does not belong to $\mathcal{A}$ anymore. This may be the case, in quantum mechanics, if the forces acting in $S$ are long-ranged.

$1, 2, \ldots, L\}$, acting on the Hilbert space $\mathcal{H}$, satisfy the CCR if

$$\left[a_l, a_n^\dagger\right] = \delta_{ln} \mathbb{1}, \qquad \left[a_l, a_n\right] = \left[a_l^\dagger, a_n^\dagger\right] = 0, \tag{2.3}$$

holds for all $l, n = 1, 2, \ldots, L$, $\mathbb{1}$ being the identity operator on $\mathcal{H}$. These operators, which are widely analyzed in any textbook on quantum mechanics (see Merzbacher, 1970; Roman, 1965), for instance, are those that are used to describe $L$ different *modes* of bosons. From these operators, we can construct $\hat{n}_l = a_l^\dagger a_l$ and $\hat{N} = \sum_{l=1}^L \hat{n}_l$, which are both self-adjoint. In particular, $\hat{n}_l$ is the *number operator* for the $l$th mode, while $\hat{N}$ is the *number operator* for $\mathcal{S}$.

An orthonormal (o.n.) basis of $\mathcal{H}$ can be constructed as follows: we introduce the *vacuum* of the theory, that is, a vector $\varphi_0$ that is annihilated by all the operators $a_l$: $a_l \varphi_0 = 0$ for all $l = 1, 2, \ldots, L$. Then we act on $\varphi_0$ with the operators $a_l^\dagger$ and with their powers

$$\varphi_{n_1, n_2, \ldots, n_L} := \frac{1}{\sqrt{n_1! \, n_2! \ldots n_L!}} (a_1^\dagger)^{n_1} (a_2^\dagger)^{n_2} \cdots (a_L^\dagger)^{n_L} \varphi_0 \tag{2.4}$$

$n_l = 0, 1, 2, \ldots$, for all $l$, and we normalize the vectors obtained in this way. The set of the $\varphi_{n_1, n_2, \ldots, n_L}$'s forms a complete and o.n. set in $\mathcal{H}$, and they are eigenstates of both $\hat{n}_l$ and $\hat{N}$:

$$\hat{n}_l \varphi_{n_1, n_2, \ldots, n_L} = n_l \varphi_{n_1, n_2, \ldots, n_L}$$

and

$$\hat{N} \varphi_{n_1, n_2, \ldots, n_L} = N \varphi_{n_1, n_2, \ldots, n_L},$$

where $N = \sum_{l=1}^L n_l$. Hence, $n_l$ and $N$ are eigenvalues of $\hat{n}_l$ and $\hat{N}$, respectively. Moreover, using the CCR we deduce that

$$\hat{n}_l \left(a_l \varphi_{n_1, n_2, \ldots, n_L}\right) = (n_l - 1)(a_l \varphi_{n_1, n_2, \ldots, n_L}),$$

for $n_l \geq 1$ whereas if $n_l = 0$, $a_l$ annihilates the vector, and

$$\hat{n}_l \left(a_l^\dagger \varphi_{n_1, n_2, \ldots, n_L}\right) = (n_l + 1)(a_l^\dagger \varphi_{n_1, n_2, \ldots, n_L}),$$

for all $l$ and for all $n_l$. For these reasons, the following interpretation is given in the literature: if the $L$ different modes of bosons of $\mathcal{S}$ are described by the vector $\varphi_{n_1, n_2, \ldots, n_L}$, this means that $n_1$ bosons are in the first mode, $n_2$ in the second mode, and so on.[2] The operator $\hat{n}_l$ acts on $\varphi_{n_1,n_2,...,n_L}$ and returns $n_l$, which is exactly the number of bosons in the $l$th mode. The operator $\hat{N}$ counts the total number of bosons. Moreover, the operator $a_l$ destroys a boson in the $l$th mode, whereas $a_l^\dagger$ creates a boson in the same mode. This is why in the physical literature $a_l$ and $a_l^\dagger$ are usually called the *annihilation* and the *creation* operators, respectively.

The vector $\varphi_{n_1,n_2,...,n_L}$ in Equation 2.4 defines a *vector (or number) state* over the set $\mathcal{A}$ as

$$\omega_{n_1,n_2,...,n_L}(X) = \langle \varphi_{n_1,n_2,...,n_L}, X\varphi_{n_1,n_2,...,n_L} \rangle, \qquad (2.5)$$

where $\langle\,,\,\rangle$ is the scalar product in the Hilbert space $\mathcal{H}$. These states are used to *project* from quantum to classical dynamics and to fix the initial conditions of the system under consideration, in a way that is clarified later. Something more concerning states is discussed later in this chapter.

For the sake of completeness, it is interesting to now check explicitly that the operators introduced so far, $a_l$, $a_l^\dagger$, $\hat{n}_l$, and $\hat{N}$, are all unbounded. This can be easily understood as, for instance,

$$\|\hat{n}_l\| = \sup_{0 \neq \varphi \in \mathcal{H}} \frac{\|\hat{n}_l \varphi\|}{\|\varphi\|} \geq \sup_{\{n_j \geq 0,\, j=1,2,...,L\}} \|\hat{n}_l \varphi_{n_1,...,n_l,...,n_L}\|$$

$$= \sup_{\{n_j \geq 0,\, j=1,2,...,L\}} n_l = \infty,$$

and it is clearly related to the fact that $\mathcal{H}$ is infinite dimensional. It is well known that unbounded operators have severe domain problems, as they cannot be defined in all of $\mathcal{H}$, (Reed and Simon, 1980), but only on a dense subset of $\mathcal{H}$. However, this is not a major problem for us here for two reasons: first, each vector $\varphi_{n_1,n_2,...,n_L}$ belongs to the domains of all the operators that are relevant to us, even when raised to some power or combined among them. Second, at least in the numerical calculations performed in Chapter 3, $\mathcal{H}$ is replaced by an *effective* Hilbert space, $\mathcal{H}_{\text{eff}}$, which becomes *dynamically* finite dimensional because of the existence of some conserved quantity and because of the initial conditions, which impose some constraints on the levels accessible to the *members* of the system. This aspect is discussed in more detail later.

---

[2]Following our preliminary discussion in Chapter 1, we could also think of $\varphi_{n_1,n_2,...,n_L}$ as the state of $L$ independent one-dimensional harmonic oscillators, each with a frequency $\omega_j$, such that the first oscillator has energy $\hbar\omega_1(n_1 + \frac{1}{2})$, the energy of the second oscillator is $\hbar\omega_2(n_2 + \frac{1}{2})$, and so on. This interpretation, however, is not interesting for our aims and is not considered in these notes.

## 2.2   THE FERMIONIC NUMBER OPERATOR

Given a set of operators $\{b_l, b_l^\dagger, \ell = 1, 2, \ldots, L\}$ acting on a certain Hilbert space $\mathcal{H}_F$, we say that they satisfy the CAR if the conditions

$$\{b_l, b_n^\dagger\} = \delta_{l,n}\mathbb{1}, \qquad \{b_l, b_n\} = \{b_l^\dagger, b_n^\dagger\} = 0 \tag{2.6}$$

hold true for all $l, n = 1, 2, \ldots, L$. Here, $\{x, y\} := xy + yx$ is the *anti-commutator* of $x$ and $y$ and $\mathbb{1}$ is now the identity operator on $\mathcal{H}_F$. These operators, which are introduced in many textbooks on quantum mechanics (Merzbacher, 1970; Roman, 1965), are those that are used to describe $L$ different *modes* of fermions. As for bosons, from these operators, we can construct $\hat{n}_l = b_l^\dagger b_l$ and $\hat{N} = \sum_{l=1}^L \hat{n}_l$, which are both self-adjoint. In particular, $\hat{n}_\ell$ is the *number operator* for the $\ell$th mode, while $\hat{N}$ is the *number operator* for $\mathcal{S}$. Compared to bosonic operators, the operators introduced in this section satisfy a very important feature: if we try to square them (or to rise to higher powers), we simply get zero: for instance, from Equation 2.6, we have $b_l^2 = 0$. As we have discussed in Chapter 1, this is related to the fact that fermions satisfy the Pauli exclusion principle, (Roman, 1965), whereas bosons do not.

The Hilbert space of our system is constructed as for bosons. We introduce the *vacuum* of the theory, that is, a vector $\Phi_0$ that is annihilated by all the operators $b_l$: $b_l \Phi_0 = 0$ for all $l = 1, 2, \ldots, L$. Then we act on $\Phi_0$ with the operators $(b_l^\dagger)^{n_l}$

$$\Phi_{n_1, n_2, \ldots, n_L} := (b_1^\dagger)^{n_1} (b_2^\dagger)^{n_2} \cdots (b_L^\dagger)^{n_L} \Phi_0, \tag{2.7}$$

$n_l = 0, 1$ for all $l$. Of course, we do not consider higher powers of the $b_j^\dagger$ as these powers would simply destroy the vector. This explains why no normalization appears. In fact, for all allowed values of the $n_l$s, the normalization constant $\sqrt{n_1! n_2! \ldots n_L!}$ in Equation 2.4 is equal to 1. These vectors form an o.n. set that spans all of $\mathcal{H}_F$ and they are eigenstates of both $\hat{n}_l$ and $\hat{N}$

$$\hat{n}_l \Phi_{n_1, n_2, \ldots, n_L} = n_l \Phi_{n_1, n_2, \ldots, n_L}$$

and

$$\hat{N} \Phi_{n_1, n_2, \ldots, n_L} = N \Phi_{n_1, n_2, \ldots, n_L},$$

where $N = \sum_{l=1}^L n_l$. A major difference with respect to what happens for bosons is that the eigenvalues of $\hat{n}_l$ are simply 0 and 1, and consequently

$N$ can take any integer value larger or equal to 0 (as for bosons). More-over, using the CAR, we deduce that

$$\hat{n}_l\left(b_l\Phi_{n_1,n_2,\dots,n_L}\right) = \begin{cases} (n_l-1)(b_l\Phi_{n_1,n_2,\dots,n_L}), & n_l = 1 \\ 0, & n_l = 0, \end{cases}$$

and

$$\hat{n}_l\left(b_l^\dagger\Phi_{n_1,n_2,\dots,n_L}\right) = \begin{cases} (n_l+1)(b_l^\dagger\Phi_{n_1,n_2,\dots,n_L}), & n_l = 0 \\ 0, & n_l = 1, \end{cases}$$

for all $l$. The interpretation does not differ from that for bosons, and then $b_l$ and $b_l^\dagger$ are again called the *annihilation* and *creation* operators, respectively. However, in some sense, $b_l^\dagger$ is **also** an annihilation operator as, acting on a state with $n_l = 1$, it destroys that state.

Of course, $\mathcal{H}_F$ has a finite dimension. In particular, for just one mode of fermions, $dim(\mathcal{H}_F) = 2$. This also implies that, contrary to what happens for bosons, all the fermionic operators are bounded and can be represented by finite-dimensional matrices.

As for bosons, the vector $\Phi_{n_1,n_2,\dots,n_L}$ in Equation 2.7 defines a *vector (or number) state* over the set $\mathcal{A}$ as

$$\omega_{n_1,n_2,\dots,n_L}(X) = \langle \Phi_{n_1,n_2,\dots,n_L}, X\Phi_{n_1,n_2,\dots,n_L}\rangle, \tag{2.8}$$

where $\langle\,,\,\rangle$ is the scalar product in $\mathcal{H}_F$. Again, these states are used to *project* from quantum to classical dynamics and to fix the initial conditions of the considered system.

**Remark:** Fermions and bosons are not the only possibilities. Other *elementary excitations*, satisfying different, and surely *less friendly*, commutation rules, are briefly considered in Chapter 10.

## 2.3 DYNAMICS FOR A QUANTUM SYSTEM

Let $S$ be a closed quantum system. This means that $S$ does not interact with any external reservoir and that its size is comparable with that of, say, the hydrogen atom: $S$ is a *microscopic* system. In this section, we describe how to find the time evolution of $S$. More precisely, we discuss in detail three possible equivalent strategies that produce, given a measurable quantity, its dependence on time.

## 2.3.1 Schrödinger Representation

The Schrödinger representation is the first way in which, in a first-level course in quantum mechanics, a student learns to describe a microscopic system $S$, using the *wave function* of $S$, $\Phi_S(t)$. This function has a probabilistic interpretation: its square modulus, $\left|\Phi_S(t)\right|^2$, gives the probability that the system $S$ is in a certain configuration corresponding to certain quantum numbers at time $t$. Just to be concrete, if we are interested in the description of the position of a particle in space, $\Phi_S(t)$ is a function of space and time, which we write here for a moment as $\Phi_S(\vec{x}; t)$, (to stress the dependence of $\Phi_S$ on the position of the particle).[3] Then $\left|\Phi_S(\vec{x}_0; t)\right|^2 d^3\vec{x}$ gives the probability of finding the particle in a region $d^3\vec{x}$ centered around the point $\vec{x}_0$, at time $t$. This probabilistic interpretation implies that, if the particle is not annihilated during its time evolution, the Hilbert norm of $\Phi_S(t)$ should stay constant in time: $\|\Phi_S(t)\| = \|\Phi_S(t_0)\|$, $t_0$ being the *initial* time. Here, if $\Phi_S(t) = \Phi_S(\vec{x}; t)$ as before, $\|\Phi_S(t)\|^2 = \int_{\mathbb{R}^3} \left|\Phi_S(\vec{x}_0; t)\right|^2 d^3\vec{x}$ is the norm in $\mathcal{L}^2(\mathbb{R}^3)$.

Following Roman (1965), it is now natural to assume that an operator $T(t, t_0)$ exists, depending on $t_0$ and on $t$, which describes the time evolution of the wave function:

$$\Phi_S(t) = T(t, t_0)\, \Phi_S(t_0). \tag{2.9}$$

It is clear that $T(t_0, t_0) = \mathbb{1}$. Moreover, as

$$\|\Phi_S(t)\|^2 = <\Phi_S(t), \Phi_S(t)> = < T(t, t_0)\, \Phi_S(t_0), T(t, t_0)\, \Phi_S(t_0)> =$$

$$= <T^\dagger(t, t_0)\, T(t, t_0)\, \Phi_S(t_0), \Phi_S(t_0)>$$

must coincide with $< \Phi_S(t_0), \Phi_S(t_0) >$, and as this has to be true for all possible choices of the initial wave function $\Phi_S(t_0)$, it follows that $T^\dagger(t, t_0)\, T(t, t_0) = \mathbb{1}$. Hence $T(t, t_0)$ is an isometric operator. As a matter of fact, $T$ is more than this: $T$ is unitary as it satisfies the equality

$$T^{-1}(t_0, t) = T(t, t_0) = T^\dagger(t_0, t) \tag{2.10}$$

---

[3]This is not necessarily the only possible choice. Many times, rather than being interested in the position of the particle, we are interested in its momentum or in its energy, for instance.

for all possible $t_0$, $t$ in $\mathbb{R}$. In fact, because of the arbitrariness of $t_0$, taken $t_1 \in [t_0, t]$, we can write $\Phi_S(t) = T(t, t_1)\, \Phi_S(t_1) = T(t, t_1)\, T(t_1, t_0)\, \Phi_S(t_0)$, so that

$$T(t, t_0) = T(t, t_1)\, T(t_1, t_0). \tag{2.11}$$

Now, taking $t \equiv t_0$ and recalling that $T(t_0, t_0) = \mathbb{1}$, we deduce that $\mathbb{1} = T(t_0, t_1)\, T(t_1, t_0)$, or $T^{-1}(t_0, t_1) = T(t_1, t_0)$. On the other hand, by left-multiplying equation $\mathbb{1} = T(t_0, t_1)\, T(t_1, t_0)$ with $T^\dagger(t_0, t_1)$, and using the equality $T^\dagger(t_0, t_1)T(t_0, t_1) = \mathbb{1}$, we deduce that $T^\dagger(t_0, t_1) = T(t_1, t_0)$. Hence, Equation 2.10 follows.

Unitarity of $T(t, t_0)$ implies (Reed and Simon, 1980) the existence of a self-adjoint operator, in general time dependent, $H(t)$, such that $T(t, t - dt) = \mathbb{1} - \frac{i}{\hbar} dt\, H(t)$. Here, $\hbar$ is used to ensure that $H(t)$ has the unit of energy.[4] Hence, as $T(t, t_0) = T(t, t - dt)\, T(t - dt, t_0)$, we get

$$\frac{1}{dt}\left(T(t, t_0) - T(t - dt, t_0)\right) = \frac{1}{dt}\left(T(t, t - dt) - \mathbb{1}\right) T(t - dt, t_0) =$$

$$= \frac{1}{dt}\left(-\frac{i}{\hbar} dt\, H(t)\right) T(t - dt, t_0) = -\frac{i}{\hbar} H(t)\, T(t - dt, t_0).$$

Taking the limit $dt \to 0$ of this equation, we deduce

$$\frac{\hbar}{i}\frac{\partial T(t, t_0)}{\partial t} + H(t)T(t, t_0) = 0, \tag{2.12}$$

which is an operatorial differential equation for $T(t, t_0)$ that must be supplemented with the initial condition $T(t_0, t_0) = \mathbb{1}$. Sometimes it may be useful, especially to set up a perturbative expansion, to rewrite Equation 2.12, with this initial condition, in the integral form

$$T(t, t_0) = \mathbb{1} - \frac{i}{\hbar}\int_{t_0}^{t} H(t')\, T(t', t_0)\, dt'. \tag{2.13}$$

Notice that, if we apply Equation 2.12 to $\Phi_S(t_0)$, we recover the well-known Schrödinger equation for the wave function $\Phi_S(t)$:

$$\frac{\hbar}{i}\frac{\partial \Phi_S(t)}{\partial t} + H(t)\, \Phi_S(t) = 0. \tag{2.14}$$

[4]Quite often in this book, we work in suitable units, taking in particular $\hbar = 1$.

In many concrete examples, the Hamiltonian does not depend explicitly on time. In this case, we can easily find the formal solutions of Equations 2.12 and 2.14:

$$
\left.
\begin{aligned}
T(t, t_0) &= \exp\left\{-\frac{i}{\hbar}(t - t_0)\,H\right\}, \\
\Phi_S(t) &= \exp\left\{-\frac{i}{\hbar}(t - t_0)\,H\right\}\Phi_S(t_0).
\end{aligned}
\right\}
\tag{2.15}
$$

The reason for calling them *formal* is that, for instance, in order to understand explicitly what $\Phi_S(t)$ really is, we have to expand the exponential, compute the action of the various terms of this expansion on $\Phi_S(t_0)$, and sum up the result. This operation, in most of the situations, cannot be performed or requires special care, and for this reason, some approximation is adopted sometimes. To be more explicit, this is often what happens when $H$ is unbounded. In fact, in this case, even if $\Phi_S(t_0)$ belongs to the domain of $H$, $D(H)$, there is no reason a priori for $\Phi_S(t_0)$ to also belong to $D(H^2)$, $D(H^3)$, and so on. Therefore, a simple-minded power expansion of $T(t, t_0)$ does not work in general.

It is important to stress that what we do observe in a concrete experiment at time $t$ is not $T(t, t_0)$ or $\Phi_S(t)$, but the *mean value* of the *observable* connected to the experiment that we are performing on $S$. For instance, if we are measuring the position of a given particle, what we measure is the mean value of the operator $\hat{r}$, the self-adjoint multiplication operator that acts on any wave function of its domain, $D(\hat{r}) \subset \mathcal{H}$, as follows: $\hat{r} f(\underline{r}) = \underline{r} f(\underline{r})$.[5]

Let $X_S$ be a certain observable, which we assume here for simplicity not to depend explicitly on time. Its mean value on the state $\Phi_S(t)$ is

$$
<X_S>_t := <\Phi_S(t), X_S\,\Phi_S(t)>. \tag{2.16}
$$

$<X_S>_t$ satisfies the following differential equation:

$$
\frac{d}{dt}<X_S>_t := <\dot{\Phi}_S(t), X_S\,\Phi_S(t)> + <\Phi_S(t), X_S\,\dot{\Phi}_S(t)> =
$$

$$
= \frac{i}{\hbar}<H(t)\Phi_S(t), X_S\Phi_S(t)> - \frac{i}{\hbar}<\Phi_S(t), X_S\,H(t)\Phi_S(t)> =
$$

---

[5]Notice that $D(\hat{r})$ cannot be all of $\mathcal{L}^2(\mathbb{R}^3)$, as it is not ensured that if $f(\underline{r}) \in \mathcal{L}^2(\mathbb{R}^3)$, $\underline{r}\,f(\underline{r})$ belongs to $\mathcal{L}^2(\mathbb{R}^3)$ as well. Nevertheless, it is not hard to prove that $D(\hat{r})$ is dense in $\mathcal{L}^2(\mathbb{R}^3)$, as it contains the space $\mathcal{S}(\mathbb{R}^3)$ of the fast decreasing $C^\infty$ functions in $\mathbb{R}^3$. The multiplication operator is *densely defined*.

$$= \frac{i}{\hbar} < \Phi_S(t), \left[H(t), X_S\right] \Phi_S(t) > = \frac{i}{\hbar} < \left[H(t), X_S\right] >_t, \qquad (2.17)$$

which follows from Equation 2.14. It is clear that the computation of $<X_S>_t$ implies the knowledge of $\Phi_S(t)$, which can be deduced solving Equation 2.14 or, using Equation 2.15, if $H$ does not depend explicitly on time. In particular, Equation 2.17 implies that if $\left[H(t), X_S\right] = 0$, the mean value $<X_S>_t$ stays constant in time.

### 2.3.2   Heisenberg Representation

The starting point to introduce the Heisenberg from the Schrödinger representation are the following definitions:

$$\left.\begin{array}{l} \Phi_H(t) := T^{-1}(t, t_0) \, \Phi_S(t) = \Phi_S(t_0), \\ X_H(t) := T^{-1}(t, t_0) \, X_S \, T(t, t_0). \end{array}\right\} \qquad (2.18)$$

In order to clarify the notation, we use the suffixes $S$ and $H$, respectively, for the Schrödinger and the Heisenberg representations (SR and HR). In the next section, we introduce a third useful way to describe the dynamics of $S$, the *interaction* representation, for which we use the suffix $I$.

The first definition in Equation 2.18, $\Phi_H(t) = \Phi_S(t_0)$ for all $t \in \mathbb{R}$, shows that the wave function of $S$ in the HR stays constant. On the contrary, the operators, in general, change with time. Indeed, the second definition in Equation 2.18 shows this dependence, which is not trivial if $\left[X_S, T(t, t_0)\right] \neq 0$. On the other hand, whenever this commutator is 0, $X_H(t) = X_S$ for all $t$. Alternatively, we can also check that $X_H(t)$ depends on $t$ just finding the differential equation, which it must satisfy. This equation follows from Equation 2.12, from its adjoint $-\frac{\hbar}{i} \frac{\partial T^\dagger(t,t_0)}{\partial t} + T^\dagger(t, t_0) \, H(t) = 0$, and from the unitarity of $T(t, t_0)$. Indeed, we have

$$\begin{aligned} \frac{dX_H(t)}{dt} &= \frac{\partial T^{-1}(t, t_0)}{\partial t} \, X_S \, T(t, t_0) + T^{-1}(t, t_0) \, X_S \, \frac{\partial T(t, t_0)}{\partial t} \\ &= \frac{i}{\hbar} T^{-1}(t, t_0) \left(H(t) X_S - X_S H(t)\right) T(t, t_0) \\ &= \frac{i}{\hbar} \left[H_H(t), X_H(t)\right], \end{aligned} \qquad (2.19)$$

where we have introduced

$$H_H(t) = T^{-1}(t, t_0) \, H \, T(t, t_0) \qquad (2.20)$$

It is important to notice that if $H$ does not depend explicitly on time, then $T(t, t_0)$ is given in Equation 2.15 and it commutes with $H$ itself, so that $H_H(t) = H$: in this case, the Hamiltonian operators in the two representations coincide. This claim is false when $H$ depends explicitly on $t$ because, in this case, $[H(t), \int_{t_0}^t H(t')dt'] \neq 0$ in general.[6]

As for the expectation values, which are now defined as

$$<X_H>_t := <\Phi_H, X_H(t)\Phi_H>, \qquad (2.21)$$

they satisfy the following differential equation

$$\frac{d}{dt}<X_H>_t := <\Phi_H, \dot{X}_H(t)\Phi_H> = \frac{i}{\hbar}<\Phi_H, \left[H_H(t), X_H(t)\right]\Phi_H>$$

$$= \frac{i}{\hbar}<[H_H(t), X_H(t)]>_t, \qquad (2.22)$$

which can be deduced by Equation 2.19. In principle, this equation is *simpler* than Equation 2.19, as it only involves mean values of operators, that is, ordinary functions of time. For instance, assuming that $H$ does not depend on time explicitly, and that $\Phi_H$ is an eigenstate of $H$, Equation 2.22 implies that $<X_H>_t$ is constant in time, even if $X_H(t)$ may show an absolutely nontrivial time evolution.

The fact that the mean values of an observable is what we really measure in an experiment suggests that they should not be depending on the representation we decide to adopt. In other words, we expect that $<X_S>_t = <X_H>_t$ for any observable $X$. Indeed we have

$$<X_S>_t = <\Phi_S(t), X_S\Phi_S(t)> = <T(t, t_0)\Phi_S(t_0), X_S\,T(t, t_0)\Phi_S(t_0)>$$

$$= <\Phi_S(t_0), T^\dagger(t, t_0)\,X_S\,T(t, t_0)\Phi_S(t_0)>$$

$$= <\Phi_H, X_H(t)\Phi_H> = <X_H>_t. \qquad (2.23)$$

### 2.3.3  Interaction Representation

The interaction representation (IR), as already stated, is a third possibility in which both the wave function and the observables of $S$ depend on time, while their mean values assume the same values as in Equation 2.23. This

---

[6]These kinds of problems are not relevant for us, as almost all the Hamiltonians we consider will not depend explicitly on time. In the only application in which the interaction Hamiltonian will depend on time, see Section 9.2, we adopt a time-dependent perturbation scheme that is described later in this chapter.

approach is useful, in particular, when the Hamiltonian of $S$ is the sum of two contributions

$$H = H_0 + H_1, \tag{2.24}$$

in which $H_0$ is the *free* Hamiltonian and $H_1$ represents the interaction between the various *components* of $S$. We could introduce the IR both from the SR or from the HR. We now briefly consider both these possibilities, starting with the first one.

**Changing Representations 1: From SR to IR**    Let $R(t_0, t)$ be a unitary operator, $R^\dagger(t_0, t) = R^{-1}(t_0, t)$, such that $R(t_0, t_0) = \mathbb{1}$ and satisfying the following differential equation:

$$\frac{\hbar}{i} \frac{\partial R(t_0, t)}{\partial t} - R(t_0, t) H_0 = 0. \tag{2.25}$$

Equation 2.25 can be rewritten as an integral equation for $R(t_0, t)$:

$$R(t_0, t) = \mathbb{1} + \frac{i}{\hbar} \int_{t_0}^{t} R(t_0, t') H_0 \, dt', \tag{2.26}$$

where the initial conditions are also taken into account. Now, given $\Phi_S(t)$ and the observable $X_S$ in the SR, we define

$$\Phi_I(t) := R(t_0, t) \, \Phi_S(t), \qquad X_I(t) := R(t_0, t) \, X_S \, R^{-1}(t_0, t). \tag{2.27}$$

Then, as $\Phi_S(t)$ must satisfy the Schrödinger equation (Eq. 2.14) with $H(t) = H_0 + H_1$, we can write

$$\frac{\partial \Phi_I(t)}{\partial t} = \frac{\partial R(t_0, t)}{\partial t} \Phi_S(t) + R(t_0, t) \frac{\partial \Phi_S(t)}{\partial t} = \frac{i}{\hbar} \left( R H_0 - R H \right) \Phi_S$$

$$= -\frac{i}{\hbar} R H_1 \Phi_S = -\frac{i}{\hbar} R H_1 R^{-1} R \Phi_S = -\frac{i}{\hbar} H_{1,I} \, \Phi_I.$$

Here we simplify the notation not writing explicitly the dependence of the operators on $t$ and $t_0$ when no confusion arises. In other words, $\Phi_I(t)$ must satisfy the equation

$$\frac{\hbar}{i} \frac{\partial \Phi_I(t)}{\partial t} + H_{1,I}(t) \, \Phi_I(t) = 0, \tag{2.28}$$

where $H_{1,I}(t) = R(t_0, t) H_1(t) R^{-1}(t_0, t)$ is the interaction Hamiltonian in IR. For what concerns the time evolution of the observables, from Equations 2.25 and 2.27 we deduce that

$$\frac{dX_I(t)}{dt} := \dot{R}(t_0, t) X_S R^{-1}(t_0, t) + R(t_0, t) X_S \dot{R}^{-1}(t_0, t)$$

$$= \frac{i}{\hbar} \left( R H_0 X_S R^{-1} - R X_S H_0 R^{-1} \right)$$

$$= \frac{i}{\hbar} \left( R H_0 R^{-1} R X_S R^{-1} - R X_S R^{-1} R H_0 R^{-1} \right)$$

as, using the unitarity of $R(t_0, t)$, we have $-\frac{\hbar}{i} \frac{\partial R^{-1}(t_0, t)}{\partial t} - H_0 R^{-1}(t_0, t)$ $= 0$. We finally get

$$\frac{dX_I(t)}{dt} = \frac{i}{\hbar} \left[ H_{0,I}(t), X_I(t) \right], \tag{2.29}$$

where $H_{0,I}(t) = R(t_0, t) H_0(t) R^{-1}(t_0, t)$. Again, if $H_0$ does not depend explicitly on $t$, $H_{0,I}$ coincides with $H_0$ as $\left[ H_0, R(t_0, t) \right] = 0$. Summarizing, we can say that *in the IR, the wave function time evolution is driven by $H_{1,I}(t)$ as in Equation* (2.28), *whereas the time dependence of the observables is governed by $H_{0,I}(t)$ as in Equation* (2.29).

**Remark:** If in the decomposition (Eq. 2.24) the interaction term is 0, $H_1 = 0$, then also $H_{1,I} = 0$ and Equation 2.28 would simply imply that the function $\Phi_I(t)$ stays constant in $t$, whereas $X_I(t)$ in general may depend on time in a nontrivial way. We are back to the HR.

**Changing Representations 2: From HR to IR**   It is possible to check that wave functions and observables in the IR and HR are related by a unitary operator $U(t, t_0)$ as follows:

$$\Phi_I(t) := U(t, t_0) \Phi_H(t), \qquad X_I(t) := U(t, t_0) X_H(t) U^{-1}(t, t_0). \tag{2.30}$$

More than this, it is also easy to relate $U(t, t_0)$ with the operators introduced so far

$$U(t, t_0) := R(t_0, t) T(t, t_0). \tag{2.31}$$

Indeed, using Equations 2.18, $\Phi_H = T^{-1} \Phi_S$ and 2.27, $\Phi_I = R \Phi_S$, and comparing what we get with the first formula in Equation 2.30, we see that

$\Phi_I = R \Phi_S = R T \Phi_H = U \Phi_H$, which must hold for all possible $\Phi_H$, so that Equation 2.31 follows. A similar computation can be performed using the second definition in Equation 2.30: we have $X_H = T^{-1} X_S T$, so that $X_S = T X_H T^{-1}$. But we also have $X_I = R X_S R^{-1} = R T X_H T^{-1} R^{-1}$, which has to be compared with $X_I = U X_H U^{-1}$, giving back, once again, Equation 2.31.

Now, the properties of $U(t, t_0)$ follow from the analogous properties of the operators $R(t_0, t)$ and $T(t, t_0)$

$$U(t_0, t_0) = R(t_0, t_0) T(t_0, t_0) = \mathbb{1} \tag{2.32}$$

Furthermore, using Equations 2.12 and 2.25

$$\frac{\partial U(t, t_0)}{\partial t} = \frac{\partial R(t, t_0)}{\partial t} T(t, t_0) + R(t, t_0) \frac{\partial T(t, t_0)}{\partial t}$$

$$= \frac{i}{\hbar} \left( R H_0 T - R H T \right) = -\frac{i}{\hbar} R H_1 T$$

$$= -\frac{i}{\hbar} R H_1 R^{-1} R T = -\frac{i}{\hbar} H_{1,I} U,$$

so that $U(t, t_0)$ has to satisfy the following differential equation:

$$\frac{\hbar}{i} \frac{\partial U(t, t_0)}{\partial t} = H_{1,I}(t) U(t, t_0) \tag{2.33}$$

In order to check that $U(t, t_0)$ is unitary, we observe that, as $R$ and $T$ are unitary,

$$U^{-1}(t, t_0) = \left( R(t_0, t) T(t, t_0) \right)^{-1} = T^{-1}(t, t_0) R^{-1}(t_0, t)$$

$$= T^\dagger(t, t_0) R^\dagger(t_0, t) = U^\dagger(t, t_0).$$

Another obvious identity is the following: $U^{-1}(t_0, t_1) = U(t_1, t_0)$, which holds for all $t_0$ and $t_1$.

As we have discussed before, what is really measured in a concrete experiment is the time evolution of the mean values of the observables. For this reason, we expect that the same evolution as the one deduced using the SR and the HR is recovered also adopting the IR. We define, in a natural way,

$$<X_I>_t := <\Phi_I(t), X_I(t) \Phi_I(t)>. \tag{2.34}$$

The first remark is that, because of Equation 2.18,

$$<X_I>_{t_0} = <\Phi_I(t_0), X_I(t_0)\,\Phi_I(t_0)>$$
$$= <\Phi_S(t_0), X_S(t_0)\,\Phi_S(t_0)> = <X_S>_{t_0}.$$

Moreover,

$$\frac{d}{dt}<X_I>_t = <\dot{\Phi}_I, X_I\,\Phi_I> + <\Phi_I, \dot{X}_I\,\Phi_I> + <\Phi_I, X_I\,\dot{\Phi}_I>,$$

where we again try to simplify the notation as much as possible. Let us now recall that $\dot{\Phi}_I = -\frac{i}{\hbar}H_{1,I}\Phi_I$, from Equation 2.28, and that $\dot{X}_I = \frac{i}{\hbar}\left[H_{0,I}, X_I\right]$, from Equation 2.29. Let us also recall that, with our rules, the scalar product is linear in the second and antilinear in the first variable.[7] Then

$$\frac{d}{dt}<X_I>_t = \frac{i}{\hbar}\left(<H_{1,I}\Phi_I, X_I\,\Phi_I> + <\Phi_I, \left[H_{0,I}, X_I\right]\Phi_I>\right.$$
$$\left.- <\Phi_I, X_I\,H_{1,I}\,\Phi_I>\right) = \frac{i}{\hbar}<\Phi_I, \left[H_I, X_I\right]\Phi_I>,$$

where $H_I = H_{0,I} + H_{1,I}$. As $\Phi_I = R\Phi_S$, $X_I = RX_SR^{-1}$ and $H_I = RHR^{-1}$, we have

$$\frac{d}{dt}<X_I>_t = \frac{i}{\hbar}<R\Phi_S, R\left[H, X_S\right]R^{-1}R\Phi_S>$$
$$= \frac{i}{\hbar}<\Phi_S, \left[H, X_S\right]\Phi_S> = \frac{d}{dt}<X_S>_t,$$

where we have used Equation 2.17 and the unitarity of $R$.

The conclusion is that $<X_I>_t$ satisfies the same differential equation as $<X_S>_t$, with the same initial condition, so that, under very general conditions, we can conclude the following:

$$<X_I>_t = <X_S>_t = <X_H>_t. \tag{2.35}$$

This means that, as expected, *the mean values of the observables and their time evolutions do not depend on the representation adopted*.

---

[7]This means that, taken $\alpha \in \mathbb{C}$ and $f, g \in \mathcal{H}$, $< f, \alpha g > = \alpha < f, g >$ while $< \alpha f, g > = \bar{\alpha} < f, g >$.

## 2.4 HEISENBERG UNCERTAINTY PRINCIPLE

Owing to its relevance in ordinary quantum mechanics, and recalling what we have already discussed in Chapter 1, we give here the mathematical details of the derivation of the Heisenberg uncertainty principle. More on this principle is discussed later in these notes, also in view of the possible applications to concrete systems.

Consider two self-adjoint noncommuting operators, $A$ and $B$, acting on the Hilbert space $\mathcal{H}$: $A = A^\dagger$, $B = B^\dagger$ and let us assume that $i\,C$ is the commutator between $A$ and $B$: $[A, B] = i\,C$. It is easy to show that, as $A = A^\dagger$ and $B = B^\dagger$, the operator $C$ must necessarily be self-adjoint as well. In fact, taking the adjoint of both sides of $[A, B] = i\,C$, we easily find that $[A, B] = i\,C^\dagger$, so that we must have $C = C^\dagger$.

Now, let $\varphi \in \mathcal{H}$ be a fixed vector for which all the mean values that we are going to compute are well defined,[8] and let us define $(\Delta A)^2 := \langle \varphi, (A - <A>)^2 \varphi \rangle = \langle A^2 \rangle - \langle A \rangle^2$, where, in general, $<X> := \langle \varphi, X\varphi \rangle$. Analogously, we have $(\Delta B)^2 = \langle B^2 \rangle - \langle B \rangle^2$.

We now observe that as each self-adjoint operator $X = X^\dagger$ has a mean value $<X>$ that is real,

$$(\Delta X)^2 = \langle \varphi, (X - <X>)^2 \varphi \rangle = \| (X - <X>)\varphi \|^2.$$

Therefore, using the Schwartz inequality,[9] we get

$$|<(A- <A>)\varphi, (B - <B>)\varphi>| \leq \|(A- <A>)\varphi\| \|(B- <B>)\varphi\|$$
$$= (\Delta A)(\Delta B).$$

Moreover, using the properties of $A$ and $<A>$, and in particular the fact that $A = A^\dagger$ and that $<A>$ is real,

$$\langle (A- <A>)\varphi, (B- <B>)\varphi \rangle = \langle \varphi, (A- <A>)(B- <B>)\varphi \rangle.$$

For our purposes, it is now useful to write

$$(A- <A>)(B- <B>)$$
$$= \frac{(A- <A>)(B- <B>) + (B- <B>)(A- <A>)}{2}$$

---

[8]This is required as $A$ and $B$ in general are unbounded.
[9]For all $f, g \in \mathcal{H}$, the Schwartz (or Cauchy–Schwarz) inequality states that $| < f, g > | \leq \|f\| \|g\|$.

$$+ \frac{(A- <A>)(B- <B>) - (B- <B>)(A- <A>)}{2}$$

$$= F + \frac{i}{2}C,$$

where we have defined a new self-adjoint operator

$$F := \frac{1}{2}[(A- <A>)(B- <B>) + (B- <B>)(A- <A>)]$$

and have observed that $(A- <A>)(B- <B>) - (B- <B>)(A- <A>) = AB - BA = i\,C$. Summarizing, we have

$$(\Delta A)^2(\Delta B)^2 \geq |\langle (A- <A>)\varphi, (B- <B>)\varphi \rangle|^2$$

$$= \left| \langle \varphi, \left( F + \frac{i}{2} C \right) \varphi \rangle \right|^2 = \left| <F> + \frac{i}{2} <C> \right|^2.$$

Now, taking into account the fact that both $<F>$ and $<C>$ are real quantities, we conclude that

$$(\Delta A)^2(\Delta B)^2 \geq \left| <F> + \frac{i}{2} <C> \right|^2 = <F>^2 + \frac{<C>^2}{4} \geq \frac{<C>^2}{4}. \tag{2.36}$$

In particular, if we now fix $A = x$, $B = p$ e $C = \mathbb{1}$, we go back to the well-known Heisenberg inequality: the commutator becomes $[x, p] = i\,\mathbb{1}$, while the inequality in Equation 2.36 takes the following well-known form: $(\Delta x)(\Delta p) \geq \frac{1}{2}$. Notice that we are working here in suitable units, taking $\hbar = 1$.

## 2.5 SOME PERTURBATION SCHEMES IN QUANTUM MECHANICS

As in classical mechanics, many quantum dynamical problems cannot be solved explicitly when the Hamiltonian is not reasonably simple. In other words, the solutions of the differential equations deduced in Section 2.3 cannot be found without considering some *smart* approximation. Hence, it turns out to be useful to apply some perturbative approach. Some of these approaches are quickly reviewed in this section, considering in particular those that are relevant for us.

## 2.5.1  A Time-Dependent Point of View

We begin our review by considering a perturbative scheme used mainly when the Hamiltonian of $S$ can be divided into a free Hamiltonian $H_0$ and an interaction part that explicitly depends on time, $\lambda\,H_I(t)$: $H(t) = H_0 + \lambda\,H_I(t)$. Here, $\lambda$ is the interaction parameter, which is usually small, or even very small, in explicit applications. We want to compute a transition probability from an initial state of $S$, $\varphi_{\mathcal{F}_0}$, to a final state $\varphi_{\mathcal{F}_f}$. We sketch this computation and refer to Messiah (1962) for more details.

To deal with this problem, it is convenient to use the SR. Hence the system is described by a time-dependent wave function $\Psi(t)$, which, for $t = 0$, reduces to $\varphi_{\mathcal{F}_0}$: $\Psi(0) = \varphi_{\mathcal{F}_0}$. The transition probability we are looking for can be written in terms of a scalar product:

$$P_{\mathcal{F}_0 \to \mathcal{F}_f}(t) := |{<}\varphi_{\mathcal{F}_f}, \Psi(t){>}|^2, \tag{2.37}$$

which is the square modulus of the projection of the state of $S$ at time $t$, $\Psi(t)$, on the *desired* final state, $\varphi_{\mathcal{F}_f}$. In the computation of this probability, we use a set of eigenvectors of the free Hamiltonian $H_0$, $\{\varphi_{\mathcal{F}}\}$, labeled by some index (or set of indexes) $\mathcal{F}$, which is an o.n. basis in the Hilbert space $\mathcal{H}$. Notice that we are assuming here that both the initial and final states of $S$ belong to this set. Then the wave function $\Psi(t)$ can be written as

$$\Psi(t) = \sum_{\mathcal{F}} c_{\mathcal{F}}(t)\, e^{-iE_{\mathcal{F}}t} \varphi_{\mathcal{F}}, \tag{2.38}$$

where $E_{\mathcal{F}}$ is the eigenvalue of $H_0$ corresponding to the eigenvector $\varphi_{\mathcal{F}}$:

$$H_0\varphi_{\mathcal{F}} = E_{\mathcal{F}}\varphi_{\mathcal{F}}, \tag{2.39}$$

and the sum is extended to all these eigenvectors. Using the quantum mechanical terminology, we could call $E_{\mathcal{F}}$ the *free energy* of $\varphi_{\mathcal{F}}$. Putting Equation 2.38 in Equation 2.37, and recalling that the eigenvectors can always be taken to satisfy the orthonormality condition $<\varphi_{\mathcal{F}}, \varphi_{\mathcal{G}}> = \delta_{\mathcal{F},\mathcal{G}}$, we have

$$P_{\mathcal{F}_0 \to \mathcal{F}_f}(t) := \left| c_{\mathcal{F}_f}(t) \right|^2. \tag{2.40}$$

Therefore, the answer to our original question is given if we are able to compute $c_{\mathcal{F}_f}(t)$ in the expansion (Eq. 2.38). In general, this computation is rather hard and cannot be carried out exactly. However, several possible perturbation schemes exist in the literature. In this section, we discuss a simple perturbation expansion in powers of the interaction parameter $\lambda$. In other words, we expand the coefficients in Equation 2.38 as follows:

$$c_{\mathcal{F}}(t) = c_{\mathcal{F}}^{(0)}(t) + \lambda c_{\mathcal{F}}^{(1)}(t) + \lambda^2 c_{\mathcal{F}}^{(2)}(t) + \cdots. \qquad (2.41)$$

The underlying idea is clear: as $\lambda$ is small, $\lambda^2$ is smaller than $\lambda$, $\lambda^3$ is even smaller, and so on. For this reason, we could imagine that, in the expansion (Eq. 2.41), only the very first contributions are relevant for us. However, to make this statement rigorous, we would also need some estimate on $|c_{\mathcal{F}}^{(j)}(t)|$, as these functions, in principle, could assume very high values during time evolution.

Each $c_{\mathcal{F}}^{(j)}(t)$ in Equation 2.41 satisfies a differential equation that can be deduced easily: first we recall that $\Psi(t)$ satisfies the Schrödinger equation $i\frac{\partial \Psi(t)}{\partial t} = H(t)\Psi(t)$. Replacing Equation 2.38 in this equation, and using the orthonormality of the vectors $\varphi_{\mathcal{F}}$'s, we find that

$$\dot{c}_{\mathcal{F}'}(t) = -i\lambda \sum_{\mathcal{F}} c_{\mathcal{F}}(t)\, e^{i(E_{\mathcal{F}'}-E_{\mathcal{F}})t} < \varphi_{\mathcal{F}'}, H_I(t)\varphi_{\mathcal{F}} >. \qquad (2.42)$$

Now, replacing Equation 2.41 in Equation 2.42, and considering equal powers of $\lambda$, we find the following infinite set of differential equations, which can be solved, in principle, up to the desired order:

$$\begin{cases} \dot{c}_{\mathcal{F}'}^{(0)}(t) = 0, \\ \dot{c}_{\mathcal{F}'}^{(1)}(t) = -i \sum_{\mathcal{F}} c_{\mathcal{F}}^{(0)}(t)\, e^{i(E_{\mathcal{F}'}-E_{\mathcal{F}})t} < \varphi_{\mathcal{F}'}, H_I(t)\varphi_{\mathcal{F}} >, \\ \dot{c}_{\mathcal{F}'}^{(2)}(t) = -i \sum_{\mathcal{F}} c_{\mathcal{F}}^{(1)}(t)\, e^{i(E_{\mathcal{F}'}-E_{\mathcal{F}})t} < \varphi_{\mathcal{F}'}, H_I(t)\varphi_{\mathcal{F}} >, \\ \cdots\cdots\cdots \end{cases} \qquad (2.43)$$

The first equation, recalling that $\Psi(0) = \varphi_{\mathcal{F}_0}$, gives $c_{\mathcal{F}'}^{(0)}(t) = c_{\mathcal{F}'}^{(0)}(0) = \delta_{\mathcal{F}',\mathcal{F}_0}$. When we replace this solution in the differential equation for $c_{\mathcal{F}'}^{(1)}(t)$, we get

$$c_{\mathcal{F}'}^{(1)}(t) = -i \int_0^t e^{i(E_{\mathcal{F}'}-E_{\mathcal{F}_0})t_1} < \varphi_{\mathcal{F}'}, H_I(t_1)\varphi_{\mathcal{F}_0} > dt_1. \qquad (2.44)$$

Using this in Equation 2.43, we further obtain

$$c_{\mathcal{F}'}^{(2)}(t) = (-i)^2 \sum_{\mathcal{F}} \int_0^t \left( \int_0^{t_2} e^{i(E_{\mathcal{F}}-E_{\mathcal{F}_0})t_1}\, h_{\mathcal{F},\mathcal{F}_0}(t_1)\, dt_1 \right)$$

$$\times\, e^{i(E_{\mathcal{F}'}-E_{\mathcal{F}})t_2}\, h_{\mathcal{F}',\mathcal{F}}(t_2)\, dt_2, \qquad (2.45)$$

where we have introduced the shorthand notation

$$h_{\mathcal{F},\mathcal{G}}(t) := <\varphi_{\mathcal{F}}, H_I(t)\varphi_{\mathcal{G}}> . \qquad (2.46)$$

Of course, this approach would produce $c_{\mathcal{F}'}^{(j)}(t)$ for any $j$, if needed. However, $\lambda$ being small, as we have already remarked only the first few $c_{\mathcal{F}'}^{(j)}(t)$s need, in general, to be computed to get a good approximation of $c_{\mathcal{F}}(t)$, and of the transition probability $P_{\mathcal{F}_0 \to \mathcal{F}_f}(t)$ as a consequence.

In the following example, we restrict this procedure to the easiest situation, that is, to a time-independent interaction Hamiltonian. In this case, the integrations can all be performed. In the applications that we consider in Chapter 9, we are interested only in those interaction Hamiltonians that are *piecewise constant*: they assume a certain value in a given time interval and, in the *next* time interval, they assume another constant value, which may, or may not, coincide with the previous one. This is used to deal with the situation in which the prices of the shares in an SSM change only when a transaction takes place and stay constant between two transactions.

**Example 2.1 Time-independent perturbation** *The first-order computation produces for $P_{\mathcal{F}_0 \to \mathcal{F}_f}(t)$ in (2.40), assuming that $\mathcal{F}_f$ is different from $\mathcal{F}_0$, the following result:*

$$P_{\mathcal{T}_0 \to \mathcal{T}_f}(t) = |c_{\mathcal{F}_f}^{(1)}(t)|^2 = \lambda^2 \left| \int_0^t e^{i(E_{\mathcal{F}_f} - E_{\mathcal{F}_0})t_1} h_{\mathcal{F}_f, \mathcal{F}_0}(t_1)\, dt_1 \right|^2, \quad (2.47)$$

*which, if $H_I(t)$ is independent of $t$ becomes*

$$P_{\mathcal{F}_0 \to \mathcal{F}_f}(t) = \lambda^2 \left( \frac{\sin(\delta Et/2)}{\delta E/2} \right)^2 |h_{\mathcal{F}_f, \mathcal{F}_0}(0)|^2. \qquad (2.48)$$

*The corresponding transition probability per unit of time turns out to be*

$$p_{\mathcal{F}_0 \to \mathcal{F}_f} := \lim_{t,\infty} \frac{1}{t} P_{\mathcal{F}_0 \to \mathcal{F}_f}(t) = 2\pi\, \lambda^2\, \delta(E_{\mathcal{F}_f} - E_{\mathcal{F}_0}) \left| h_{\mathcal{F}_f, \mathcal{F}_0}(0) \right|^2. \qquad (2.49)$$

*This formula shows that, in this limit, a transition between two states is possible only if the two states have the same free energy. The presence of $h_{\mathcal{F}_f, \mathcal{F}_0}(0)$ in the final result shows that a transition is allowed only if the matrix elements of $H_I$ in our basis, $h_{\mathcal{F},\mathcal{G}} := <\varphi_{\mathcal{F}}, H_I \varphi_{\mathcal{G}}>$, are not all*

*zero. The other transitions are forbidden. In particular, $< \varphi_{\mathcal{F}_f}, H_I \varphi_{\mathcal{F}_0} >$
must be different from zero here, to get a nontrivial value for $p_{\mathcal{F}_0 \to \mathcal{F}_f}$.*
*Looking for a better approximation, Equation 2.45 produces*

$$c_{\mathcal{F}_f}^{(2)}(t) = \sum_{\mathcal{F}} h_{\mathcal{F}_f, \mathcal{F}}(0) h_{\mathcal{F}, \mathcal{F}_0}(0) \mathcal{E}_{\mathcal{F}, \mathcal{F}_0, \mathcal{F}_f}(t), \qquad (2.50)$$

*where*

$$\mathcal{E}_{\mathcal{F}, \mathcal{F}_0, \mathcal{F}_f}(t) = \frac{1}{E_{\mathcal{F}} - E_{\mathcal{F}_0}} \left( \frac{e^{i(E_{\mathcal{F}_f} - E_{\mathcal{F}_0})t} - 1}{E_{\mathcal{F}_f} - E_{\mathcal{F}_0}} - \frac{e^{i(E_{\mathcal{F}_f} - E_{\mathcal{F}})t} - 1}{E_{\mathcal{F}_f} - E_{\mathcal{F}}} \right)$$

*Recalling definition in Equation 2.46, we rewrite Equation 2.50 as*
$c_{\mathcal{F}_f}^{(2)}(t) = \sum_{\mathcal{F}} < \varphi_{\mathcal{F}_f}, H_I \varphi_{\mathcal{F}} > < \varphi_{\mathcal{F}}, H_I \varphi_{\mathcal{F}_0} > \mathcal{E}_{\mathcal{F}, \mathcal{F}_0, \mathcal{F}_f}(t)$. *This formula*
*shows that for the second-order corrections of $c_{\mathcal{F}_f}(t)$ to be nonzero, it*
*is enough that some intermediate $\varphi_{\mathcal{F}}$ exists such that the corresponding*
*matrix elements of $H_I$ are different from zero.*

We refer to Bagarello (2009) for an extension of this result to the case
in which the interaction Hamiltonian changes discontinuously in time few
times. Concrete applications of this procedure are considered in some
detail in Examples 9.1 and 9.2, in connection with our stock markets.

## 2.5.2 Feynman Graphs

Following (Messiah, 1962), we now try to connect the analytic expression
of a given approximation of $c_{\mathcal{F}_f}(t)$ with some kind of Feynman graph in
such a way that the higher orders could be easily written considering a
certain set of rules that we call, for obvious reasons, *Feynman rules*.
    The starting point is given by the expressions in Equations 2.44 and
2.45 for $c_{\mathcal{F}_f}^{(1)}(t)$ and $c_{\mathcal{F}_f}^{(2)}(t)$, which is convenient to rewrite in the following
form:

$$c_{\mathcal{F}_f}^{(1)}(t) = -i \int_0^t e^{i E_{\mathcal{F}_f} t_1} < \varphi_{\mathcal{F}_f}, H_I(t_1) \varphi_{\mathcal{F}_0} > e^{-i E_{\mathcal{F}_0} t_1} \, dt_1, \qquad (2.51)$$

and

$$c_{\mathcal{F}_f}^{(2)}(t) = (-i)^2 \sum_{\mathcal{F}} \int_0^t dt_2 \int_0^{t_2} dt_1 \, e^{i E_{\mathcal{F}_f} t_2} < \varphi_{\mathcal{F}_f}, H_I(t_2) \varphi_{\mathcal{F}} > e^{-i E_{\mathcal{F}} t_2}$$

$$\times e^{i E_{\mathcal{F}} t_1} < \varphi_{\mathcal{F}}, H_I(t_1) \varphi_{\mathcal{F}_0} > e^{-i E_{\mathcal{F}_0} t_1}. \qquad (2.52)$$

A graphical way to describe $c_{\mathcal{F}_f}^{(1)}(t)$ in Equation 2.51 is given in Figure 2.1: at $t = t_0$ the state of the system is $\varphi_{\mathcal{F}_0}$, which evolves freely (and therefore $e^{-iE_{\mathcal{F}_0}t_1}\varphi_{\mathcal{F}_0}$ appears in the integral) until the interaction occurs, at $t = t_1$. After the interaction, the system is moved to the state $\varphi_{\mathcal{F}_f}$, which, for $t > t_1$, evolves again freely (therefore $e^{-iE_{\mathcal{F}_f}t_1}\varphi_{\mathcal{F}_f}$ appears, and the different sign in Equation 2.51 is due to the antilinearity of the scalar product in the first variable). The free evolutions are the upward inclined arrows, while the interaction between the initial and the final states, $< \varphi_{\mathcal{F}_f}, H_I(t_1)\varphi_{\mathcal{F}_0} >$, is described by the horizontal wavy line in the figure. As the interaction may occur at any time between 0 and $t$, we have to integrate on all these possible $t_1$s. We finally multiply the result by $-i$, which can be considered as a sort of normalization constant.

In a similar way, we can construct the Feynman graph for $c_{\mathcal{F}_f}^{(2)}(t)$, $c_{\mathcal{F}_f}^{(3)}(t)$ and so on. For example, $c_{\mathcal{F}_f}^{(2)}(t)$ in Equation 2.52 can be deduced by a graph such as the one in Figure 2.2, where two interactions occur: the first at $t = t_1$ and the second at $t = t_2 > t_1$.

Because of the double interaction, we have to integrate the result twice, recalling that $t_1 \in (0, t_2)$ and $t_2 \in (0, t)$. For the same reason, we have to sum over all the possible intermediate states, $\varphi_{\mathcal{F}}$. The free time evolutions for the various free fields also appear, as well as an overall factor $(-i)^2$, one $-i$ for each integration. Following these same rules, we could also give at least a formal expression for the other coefficients, as $c_{\mathcal{F}_f}^{(3)}(t)$, $c_{\mathcal{F}_f}^{(4)}(t)$, and so on. The third-order correction $c_{\mathcal{F}_f}^{(3)}(t)$ contains, for instance, a double sum on the intermediate states, a triple time integration, and a normalization $(-i)^3$. Of course, this approach is, up to this stage, purely formal as all the integrals should be explicitly computed, and this may

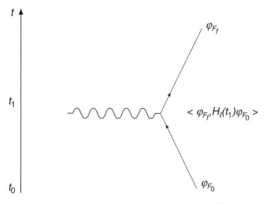

**Figure 2.1**   Graphical expression for $c_{\mathcal{F}_f}^{(1)}(t)$.

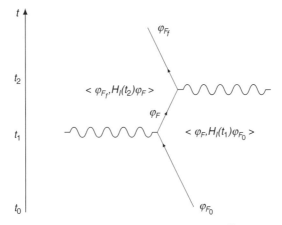

**Figure 2.2**  Graphical expression for $c^{(2)}_{\mathcal{F}_f}(t)$.

be a nontrivial task, also because the range of the integrations are all different.

### 2.5.3  Dyson's Perturbation Theory

This last problem can be formally solved by the technique that we review briefly here in a slightly different context, that is, in an attempt to find an explicit expression for the operator $U(t, t_0)$ introduced in Section 2.3.3. The starting point is Equation 2.33, which, together with the initial condition $U(t_0, t_0) = \mathbb{1}$, can be rewritten in integral form as

$$U(t, t_0) = \mathbb{1} - \frac{i}{\hbar} \int_{t_0}^{t} H_{1,I}(\tau_1) \, U(\tau_1, t_0) \, d\tau_1. \qquad (2.53)$$

Quite often, in concrete applications, the operator $H_1$, and $H_{1,I}$ as a consequence, depend on a small parameter, $\lambda$. This is when $H_1$ is a perturbation of the free Hamiltonian. If this is the case, it is natural to look for approximated solutions for $U(t, t_0)$, and this can be achieved expanding Equation 2.53 in powers of $\lambda$. At the zeroth order in $\lambda$, we have

$$U(t, t_0) \simeq U^{(0)}(t, t_0) = \mathbb{1}.$$

At the first order, which we obtain inserting $U^{(0)}(t, t_0)$ in the right-hand side of Equation 2.53, we get

$$U(t, t_0) \simeq U^{(1)}(t, t_0) := \mathbb{1} - \frac{i}{\hbar} \int_{t_0}^{t} H_{1,I}(\tau_1) \, d\tau_1.$$

In the same way, the second order turns out to be

$$U(t, t_0) \simeq U^{(2)}(t, t_0) = \mathbb{1} - \frac{i}{\hbar} \int_{t_0}^{t} H_{1,I}(\tau_1) U^{(1)}(\tau_1, t_0) \, d\tau_1$$

$$= \mathbb{1} - \frac{i}{\hbar} \int_{t_0}^{t} H_{1,I}(\tau_1) \, d\tau_1 + \left(-\frac{i}{\hbar}\right)^2 \int_{t_0}^{t} H_{1,I}(\tau_1)$$

$$\times \left(\int_{t_0}^{\tau_1} H_{1,I}(\tau_2) \, d\tau_2\right) d\tau_1.$$

The procedure can be iterated, and the result is the following infinite series:

$$U(t, t_0) = \sum_{n=0}^{\infty} U_n(t, t_0), \qquad (2.54)$$

where

$$U_n(t, t_0) = \left(-\frac{i}{\hbar}\right)^n \int_{t_0}^{t} \int_{t_0}^{\tau_1} \cdots \int_{t_0}^{\tau_{n-1}} H_{1,I}(\tau_1)$$

$$\times H_{1,I}(\tau_2) \cdots H_{1,I}(\tau_n) \, d\tau_n \cdots d\tau_2 \, d\tau_1. \qquad (2.55)$$

**Remark:** It is important to stress that in the computation of $U_n(t, t_0)$, both the integration order (we have to first compute the integral in $d\tau_n$, then the one in $d\tau_{n-1}$, and so on) and the position of the various operators $H_{1,I}(\tau_j)$ (which, in general, do not commute), are crucial and should be properly taken into account. Recall also that the various $\tau_j$ are ordered as follows:

$$t_0 \leq \tau_n \leq \tau_{n-1} \leq \cdots \leq \tau_1 \leq t. \qquad (2.56)$$

One of the main difficulties in computing explicitly the integral in Equation 2.55 is that all the integration ranges are different. This is exactly the same problem found in the computation of the various contributions $c_{\mathcal{F}}^{(n)}(t)$ in Equation 2.41, and for this reason, it is not surprising that the same technique can be applied to both these situations. An easy (but formal) solution can be obtained by introducing the Dyson's operator of *chronological order*, $P$, which acts on two generic time-dependent operators $A_1(t)$ and $A_2(t)$ as follows:

$$P\left(A_1(\tau_1)A_2(\tau_2)\right) = \begin{cases} A_1(\tau_1)A_2(\tau_2), & \text{if } \tau_1 > \tau_2; \\ A_2(\tau_2)A_1(\tau_1), & \text{if } \tau_1 < \tau_2. \end{cases} \qquad (2.57)$$

In other words, $P$ acts on certain time-dependent operators ordering them in chronological order from the right to the left. Then, if $\tau_1 > \tau_2 > \cdots > \tau_n$, we get

$$P\left(A_i(\tau_i)A_j(\tau_j)\cdots A_l(\tau_l)\right) = A_1(\tau_1)A_2(\tau_2)\cdots A_n(\tau_n). \tag{2.58}$$

It is clear that an ambiguity appears whenever $\tau_j = \tau_i$, for some $i$ and $j$. There is no ambiguity, however, if we also have $A_j = A_i$, which is exactly the case we are interested in here.[10] In this case, we get

$$U_n(t, t_0) = \frac{1}{n!} \left(-\frac{i}{\hbar}\right)^n \int_{t_0}^t \int_{t_0}^t \cdots$$
$$\int_{t_0}^t P\left(H_{1,I}(\tau_1)\, H_{1,I}(\tau_2) \cdots H_{1,I}(\tau_n)\right) d\tau_n \cdots d\tau_2\, d\tau_1,$$
$$\tag{2.59}$$

(Roman, 1965), and then we can formally sum up the series in Equation 2.54 finding

$$U(t, t_0) = P\left\{\exp\left\{-\frac{i}{\hbar}\int_{t_0}^t H_{1,I}(\tau)\, d\tau\right\}\right\}, \tag{2.60}$$

which is the solution of our differential equation (Eq. 2.33). The operator $P$ could also be used in the computation of the various terms $c_{\mathcal{F}}^{(n)}(t)$ in Equation 2.41.

### 2.5.4 The Stochastic Limit

Another perturbative scheme that we use later in this book is the so-called stochastic limit approach (SLA). In this section, we briefly summarize some of its basic facts and properties. However, the reader should be aware of the fact that the SLA is a very difficult subject and a complete review would take much more than needed in these notes. For this reason, we limit here to suggest a list of rules useful to achieve our final aim, which is essentially the deduction of a simplified version of the original dynamics for the model considered. We refer to Accardi et al. (2002) and references therein for (many) more details.

Given an open system $\mathcal{S} \cup \mathcal{R}$, $\mathcal{S}$ being the *main system* and $\mathcal{R}$ the reservoir coupled to $\mathcal{S}$, we write its Hamiltonian $H$ as the sum of two contributions, the free part $H_0$ and the interaction $\lambda H_I$. Here, $\lambda$ is a coupling

---

[10]In our computations, all the operators $A_j$ coincide with the Hamiltonian $H$ of $\mathcal{S}$.

constant, $H_0$ contains the free evolution of both the system $S$ and the reservoir $R$, and $H_I$ describes the interaction between $S$ and $R$. Working in the interaction picture, see Section 2.3.3, and adopting a slightly simplified notation, we define $H_I(t) = e^{iH_0t} H_I e^{-iH_0t}$ and the so-called wave operator $U_\lambda(t)$, which is the solution of the following differential equation:

$$\partial_t U_\lambda(t) = -i\lambda H_I(t)U_\lambda(t), \tag{2.61}$$

with the initial condition $U_\lambda(0) = \mathbb{1}$. Using the van Hove rescaling $t \to \frac{t}{\lambda^2}$ (see Accardi et al., 2002), we can rewrite the same equation in a form that is more convenient for our perturbative approach, that is,

$$\partial_t U_\lambda\left(\frac{t}{\lambda^2}\right) = -\frac{i}{\lambda}H_I\left(\frac{t}{\lambda^2}\right)U_\lambda\left(\frac{t}{\lambda^2}\right), \tag{2.62}$$

with the same initial condition as before. Its integral counterpart looks like

$$U_\lambda\left(\frac{t}{\lambda^2}\right) = \mathbb{1} - \frac{i}{\lambda}\int_0^t H_I\left(\frac{t'}{\lambda^2}\right)U_\lambda\left(\frac{t'}{\lambda^2}\right) dt', \tag{2.63}$$

which is the starting point for a perturbative expansion, which works in the following way: let $\varphi_0$ be the ground vector of the reservoir, and $\xi$ a generic vector of the systems. We put $\varphi_0^{(\xi)} = \varphi_0 \otimes \xi$, which is a vector of $S \cup R$. Incidentally, we observe that what we are considering here is a zero-temperature reservoir. If we want to consider different reservoirs, and insert the temperature in the game, we should replace the state generated by the vector $\varphi_0$ with a KMS state[11] (Bratteli and Robinson, 1987a,b; Sewell, 1989, 2002), that is, with a thermal equilibrium state. Some aspects of these and other states in quantum mechanics are reviewed in the next section.

Going back to our original problem, we want to compute the limit, for $\lambda$ going to 0, of the first nontrivial order of the mean value of the perturbative expansion of $U_\lambda(t/\lambda^2)$ above in the vector state defined by $\varphi_0^{(\xi)}$, that is, the limit of

$$I_\lambda(t) = \left(-\frac{i}{\lambda}\right)^2 \int_0^t dt_1 \int_0^{t_1} dt_2 \left\langle H_I\left(\frac{t_1}{\lambda^2}\right) H_I\left(\frac{t_2}{\lambda^2}\right)\right\rangle_{\varphi_0^{(\xi)}}, \tag{2.64}$$

[11]KMS are the initials of Ryogo Kubo, Paul C. Martin, and Julian S. Schwinger.

for $\lambda \to 0$. Under some regularity conditions on the functions that are used to smear out the (typically) bosonic fields of the reservoir, this limit is shown to exist for many physically relevant models (see Accardi et al., 2002 and Bagarello, 2005). We define $I(t) := \lim_{\lambda \to 0} I_\lambda(t)$. In the same sense of the convergence of the (rescaled) wave operator $U_\lambda(\frac{t}{\lambda^2})$ (the convergence in the sense of correlators), it is possible to check that the (rescaled) reservoir operators also converge and define new operators that do not satisfy CCR but a modified version of these (Bagarello, 2005). Moreover, these limiting operators depend explicitly on time, and they *live* in a Hilbert space that is different from the original one. In particular, they annihilate a vacuum vector, $\eta_0$, which is no longer the original one, $\varphi_0$.

It is not difficult, now, to deduce the form of a time-dependent self-adjoint operator $H_I^{(sl)}(t)$, which depends on the system operators and on the limiting operators of the reservoir, such that the first nontrivial order of the mean value of the expansion of

$$ U_t := \mathbb{1} - i \int_0^t H_I^{(sl)}(t') U_{t'} \mathrm{d}t' $$

on the state $\eta_0^{(\xi)} = \eta_0 \otimes \xi$ coincides with $I(t)$. The operator $U_t$ defined by this integral equation is called the *wave operator*.

The form of the generator of the *reduced* dynamics now follows from an operation of normal ordering. In more detail, we start defining the flux of an observable $\tilde{X} = X \otimes \mathbb{1}_R$, where $\mathbb{1}_R$ is the identity of the reservoir and $X$ is an observable of the system, as $j_t(\tilde{X}) := U_t^\dagger \tilde{X} U_t$. Then, using the equation of motion for $U_t$ and $U_t^\dagger$, we find that $\partial_t j_t(\tilde{X}) = iU_t^\dagger \left[ H_I^{(sl)}(t), \tilde{X} \right] U_t$. In order to compute the mean value of this equation on the state $\eta_0^{(\xi)}$, and so to get rid of the reservoir operators, it is convenient to compute first the commutation relations between $U_t$ and the limiting operators of the reservoir. At this stage, the so-called time consecutive principle is used in a very heavy way to simplify the computation. This principle, which has been checked for many classes of physical models, (Accardi et al., 2002), states that, if $\beta(t)$ is any of these limiting operators of the reservoir, then

$$ \left[ \beta(t), U_{t'} \right] = 0, \tag{2.65} $$

for all $t > t'$. Using this principle and recalling that $\eta_0$ is annihilated by the limiting annihilation operators of the reservoir, see Section 8.3.1 for an explicit example, it is now a simple exercise to compute $\langle \partial_t j_t(\tilde{X}) \rangle_{\eta_0^{(\xi)}}$

and, by means of the equation $\langle \partial_t j_t(\tilde{X}) \rangle_{\eta_0^{(\xi)}} = \langle j_t(L(\tilde{X})) \rangle_{\eta_0^{(\xi)}}$, to identify the form of the generator $L$ of the physical system. This allows us to obtain equations of motion that are, in general, much easier than the original ones, as the contribution of the reservoir has been, in a certain sense, already taken into account during the limiting procedure. We refer to Bagarello (2005) for a review of some applications of the SLA to quantum many-body systems.

## 2.6   FEW WORDS ON STATES

In general, a state $\omega$ over a certain algebra $\mathcal{A}$ is a positive linear functional that is normalized. In other words, for each bounded operator $A$, $\omega$ satisfies the following:

$$\omega(A^\dagger A) \geq 0, \quad \text{and} \quad \omega(\mathbb{1}) = 1.$$

Moreover, $\omega(A^\dagger) = \overline{\omega(A)}$, for all $A \in \mathcal{A}$. States over bounded operators share many good properties: they are automatically continuous (Bratteli and Robinson, 1987): if $A_n$ is a sequence of bounded operators converging to $A$ in the uniform topology, then $\omega(A_n)$ converges to $\omega(A)$ in the topology of $\mathbb{C}$. A very important inequality that each state satisfies is the *Cauchy–Schwarz inequality* already considered, in a slightly different form, in Section 2.4: for each $A$ and $B$ bounded, the following holds:

$$|\omega(A^\dagger B)|^2 \leq \omega(A^\dagger A)\omega(B^\dagger B).$$

The states introduced so far (Eqs 2.5 and 2.8) describe a situation in which the numbers of all the different modes of bosons or fermions are, in a sense, known. These are particular examples of the so-called *vector states*: taking a normalized vector $\varphi \in \mathcal{H}$ and defining a map from the algebra of the bounded operators $B(\mathcal{H})$ into $\mathbb{C}$ as

$$B(\mathcal{H}) \ni A \longrightarrow \Omega_\varphi(A) := \langle \varphi, A \varphi \rangle,$$

it turns out that $\Omega_\varphi$ is a state over $B(\mathcal{H})$. But other kind of states also exist and are relevant in many physical applications. In particular, the previously cited KMS state, that is, the equilibrium states for systems with infinite degrees of freedom, are usually used to prove the existence of phase transitions or to find conditions for some thermodynamical equilibrium to exist. Without going into the rigorous mathematical definition, see

Bratteli and Robinson, 1987, a KMS state $\omega$ with inverse temperature $\beta$ satisfies the following equality, known as *the KMS condition*:

$$\omega(AB(i\beta)) = \omega(BA), \qquad (2.66)$$

where $A$ and $B$ are general elements of $B(\mathcal{H})$ and $B(i\beta)$ is the time evolution of the operator $B$ computed at the complex value $i\beta$ of the time. It is well known that (Sewell, 1989), when restricted to a finite size system, a KMS state is nothing but a Gibbs state. In physical applications, a KMS state is used to describe a thermal bath interacting with a physical system $\mathcal{S}$, if the bath has a nonzero temperature.

## 2.7   GETTING AN EXPONENTIAL LAW FROM A HAMILTONIAN

If $\mathcal{S}$ is a closed quantum system with time-independent self-adjoint Hamiltonian $H$, it is natural to suspect that only periodic or quasi-periodic effects can take place, during the time evolution of $\mathcal{S}$. This is because the energy of $\mathcal{S}$ is preserved, and this seems to prevent any damping effect for $\mathcal{S}$. For instance, if we work in the SR, the time evolution $\Psi(t)$ of the wave function of the system is simply $\Psi(t) = e^{-iHt}\Psi(0)$ and, as the operator $e^{-iHt}$ is unitary, we do not expect that $\Psi(t)$ decreases (in some suitable sense) to 0 when $t$ diverges. Nevertheless, we show that a similar decay feature is possible if $\mathcal{S}$ is coupled to a reservoir $\mathcal{R}$, but only if $\mathcal{R}$ is rather *large* compared to $\mathcal{S}$, or, more explicitly, if $\mathcal{R}$ has an infinite number of degrees of freedom.

We start considering a first system, $\mathcal{S}$, interacting with a second system, $\tilde{\mathcal{S}}$, and we assume for the time being that both $\mathcal{S}$ and $\tilde{\mathcal{S}}$ are *of the same size*: to be concrete, this means here that $\mathcal{S}$ describes a single particle whose related operators are $a$, $a^\dagger$, and $\hat{n}_a = a^\dagger a$ and, analogously, $\tilde{\mathcal{S}}$ describes a second particle whose related operators are $b$, $b^\dagger$, and $\hat{n}_b = b^\dagger b$. These operators obey the following CCR: $[a, a^\dagger] = [b, b^\dagger] = \mathbb{1}$, whereas all the other commutators are assumed to be 0. A natural choice for the Hamiltonian of $\mathcal{S} \cup \tilde{\mathcal{S}}$ is the following: $h = \omega_a \hat{n}_a + \omega_b \hat{n}_b + \mu\left(a^\dagger b + b^\dagger a\right)$, where $\omega_a$, $\omega_b$ and $\mu$ must be real quantities for $h$ to be self-adjoint. Recall that losing self-adjointness of $h$ would produce a nonunitary time evolution, and this is out of the scheme, usually considered in ordinary quantum mechanics.[12] The Hamiltonian $h$ contains a free part plus an interaction,

---

[12]We should also mention that in recent years, larger and larger groups of physicists and mathematicians have started to become interested in what happens when the dynamics is defined by means of an operator that is not self-adjoint but satisfies certain symmetry properties that are physically motivated. This is what, in the literature, is usually called *pseudo-Hermitian quantum mechanics*; see also Section 2.7.1.

which is such that if the eigenvalue of $\hat{n}_a$ increases by 1 unit during the time evolution, the eigenvalue of $\hat{n}_b$ must decrease by 1 unit, and vice versa. This is because $[h, \hat{n}_a + \hat{n}_b] = 0$, so that $\hat{n}_a + \hat{n}_b$ is an integral of motion. The equations of motion for $a(t)$ and $b(t)$ can be easily deduced and turn out to be

$$\dot{a}(t) = i\,[h, a(t)] = -i\omega_a a(t) - i\mu b(t),$$

$$\dot{b}(t) = i\,[h, b(t)] = -i\omega_b b(t) - i\mu a(t),$$

whose solution can be written as $a(t) = a\alpha_a(t) + b\alpha_b(t)$ and $b(t) = a\beta_a(t) + b\beta_b(t)$, where the functions $\alpha_j(t)$ and $\beta_j(t)$, $j = a, b$, are linear combinations of $e^{\lambda_\pm t}$, with $\lambda_\pm = \frac{-i}{2}(\omega_a + \omega_b - \sqrt{(\omega_a - \omega_b)^2 + 4\mu^2})$. Moreover, $\alpha_a(0) = \beta_b(0) = 1$ and $\alpha_b(0) = \beta_a(0) = 0$, in order to have $a(0) = a$ and $b(0) = b$. Hence we see that both $a(t)$ and $b(t)$, and $\hat{n}_a(t) = a^\dagger(t)a(t)$ and $\hat{n}_b(t) = b^\dagger(t)b(t)$ as a consequence, are linear combinations of oscillating functions so that no damping is possible within this simple model. Incidentally, we observe that we have adopted here the HR in the computation of the dynamics, as we quite often do in the rest of this book.

Suppose now that the system $\tilde{\mathcal{S}}$ is replaced by an (infinitely extended) reservoir $\mathcal{R}$, whose particles are described by an infinite set of bosonic operators $b(k)$, $b^\dagger(k)$, and $\hat{n}(k) = b^\dagger(k)b(k)$, $k \in \mathbb{R}$. The Hamiltonian of $\mathcal{S} \cup \mathcal{R}$ extends $h$ above and is now taken to be

$$H = H_0 + \lambda H_I, \quad H_0 = \omega \hat{n}_a + \int_{\mathbb{R}} \omega(k)\hat{n}(k)\,dk,$$

$$H_I = \int_{\mathbb{R}} \left(ab^\dagger(k) + a^\dagger b(k)\right) f(k)dk, \tag{2.67}$$

where $[a, a^\dagger] = \mathbb{1}$, $[b(k), b^\dagger(q)] = \mathbb{1}\delta(k - q)$, while all the other commutators are 0. All the constants appearing in Equation 2.67, as well as the regularizing function $f(k)$, are real, so that $H = H^\dagger$. At a certain point in this section, we take $f(k) \equiv 1$: this makes the computation of the solution of the equations of motion easier, but makes our $H$ a formal object. A more rigorous approach could be settled using the results in Accardi et al. (2002). However, most of the times we do not consider this possibility in these notes, as our main interest is in finding a reasonably simple technique, rather than focusing a lot on the mathematical rigor. Notice that an integral of motion also exists for $\mathcal{S} \cup \mathcal{R}$, $\hat{n}_a + \int_{\mathbb{R}} \hat{n}(k)\,dk$, which extends the one for $\mathcal{S} \cup \tilde{\mathcal{S}}$, $\hat{n}_a + \hat{n}_b$. With this choice of $H$, the Heisenberg equations of motions are

$$\begin{cases} \dot{a}(t) = i\,[H, a(t)] = -i\omega a(t) - i\lambda \int_{\mathbb{R}} f(k)\,b(k,t)\,dk, \\ \dot{b}(k,t) = i\,[H, b(k,t)] = -i\omega(k)b(k,t) - i\lambda f(k)\,a(t), \end{cases} \qquad (2.68)$$

which are supplemented by the initial conditions $a(0) = a$ and $b(k,0) = b(k)$. The last equation can be rewritten as $b(k,t) = b(k)e^{-i\omega(k)t} - i\lambda f(k) \int_0^t a(t_1)e^{-i\omega(k)(t-t_1)}\,dt_1$. Now fixing $f(k) = 1$ and choosing $\omega(k) = k$, replacing $b(k,t)$ in the first line in Equation 2.68, with a change of the order of integration, and recalling that $\int_{\mathbb{R}} e^{-ik(t-t_1)}\,dk = 2\pi\delta(t-t_1)$ and that $\int_0^t g(t_1)\delta(t-t_1)\,dt_1 = \frac{1}{2}g(t)$ for any test function $g(t)$, we conclude that

$$\dot{a}(t) = -(i\omega + \pi\lambda^2)a(t) - i\lambda \int_{\mathbb{R}} b(k)\,e^{-ikt}\,dk.$$

This equation can be solved, and the solution can be written as

$$a(t) = \left(a - i\lambda \int_{\mathbb{R}} dk\,\eta(k,t)b(k)\right) e^{-(i\omega + \pi\lambda^2)t}, \qquad (2.69)$$

where $\eta(k,t) = \frac{1}{\rho(k)}\left(e^{\rho(k)t} - 1\right)$ and $\rho(k) = i(\omega - k) + \pi\lambda^2$. Using complex contour integration, it is possible to check that $[a(t), a^\dagger(t)] = \mathbb{1}$ for all $t$: this means that the natural decay of $a(t)$, described in (2.69), is balanced by the reservoir contribution. This feature is crucial as it is a measure of the fact that the time evolution is unitarily implemented, even if $a(t)$ decays for $t$ increasing.

Let us now consider a state over $\mathcal{S} \cup \mathcal{R}$, $\langle X_\mathcal{S} \otimes X_\mathcal{R} \rangle = \langle \varphi_{n_a}, X_\mathcal{S}\varphi_{n_a} \rangle \langle X_\mathcal{R} \rangle_\mathcal{R}$, in which $\varphi_{n_a}$ is the eigenstate of the number operator $\hat{n}_a$ and $< >_\mathcal{R}$ is a state of the reservoir, see Section 2.6, which is assumed to satisfy $\langle b^\dagger(k)b(q) \rangle_\mathcal{R} = n_b(k)\delta(k-q)$. This is a standard choice (see for instance Barnett and Radmore, 1997), which extends the choice we made for $\mathcal{S}$. Here, $X_\mathcal{S} \otimes X_\mathcal{R}$ is the tensor product of an operator of the system, $X_\mathcal{S}$, and an operator of the reservoir, $X_\mathcal{R}$. Then, if for simplicity we take the function $n_b(k)$ to be constant in $k$, we get calling $n_a(t) := < \hat{n}_a(t) > = < a^\dagger(t)a(t) >$,

$$n_a(t) = n_a\,e^{-2\lambda^2\pi t} + n_b\left(1 - e^{-2\lambda^2\pi t}\right), \qquad (2.70)$$

which goes to $n_b$ as $t \to \infty$. Hence, if $0 \le n_b < n_a$, the value of $n_a(t)$ decreases with time. If, on the other hand, $n_b > n_a$, the value of $n_a(t)$

increases for large $t$. This is the exponential rule that, as discussed before, cannot be deduced if $\mathcal{R}$ does not have an infinite number of degrees of freedom. Notice that, in particular, if the reservoir is originally empty, $n_b = 0$, then $n_a(t) = n_a\, e^{-2\lambda^2\pi t}$ decreases exponentially to zero: the system becomes empty. On the other hand, as $\hat{n}_a + \int_{\mathbb{R}} \hat{n}(k)\, dk$ is a constant of motion, the reservoir starts to be filled up.

It might be interesting to note that the *continuous* reservoir considered here could be replaced by a discrete one, describing again an infinite number of particles, but labeled by a discrete index. In this case, to obtain a Dirac delta distribution, which is the crucial ingredient in the derivation above, we have to replace the integral $\int_{\mathbb{R}} e^{-ik(t-t_1)}\, dk = 2\pi\delta(t - t_1)$ with the *Poisson summation formula*, which we write here as $\sum_{n\in\mathbb{Z}} e^{inxc} = \frac{2\pi}{|c|} \sum_{n\in\mathbb{Z}} \delta\left(x - n\frac{2\pi}{c}\right)$, for all possible nonzero $c \in \mathbb{R}$.

### 2.7.1   Non-Self-Adjoint Hamiltonians for Damping

A simple-minded point of view on damping effects might suggest that in an attempt to describe these kinds of effects, it could be convenient to use non-self-adjoint Hamiltonians to deduce the dynamics and that there is no need to consider any reservoir interacting with the system $\mathcal{S}$. This is, indeed, the point of view of many papers in quantum optics. Just to cite a simple example (Ben-Aryeh et al., 2004; Cherbal et al., 2007), an effective non-self-adjoint Hamiltonian describing a two-level atom interacting with an electromagnetic field was analyzed in connection with pseudo-Hermitian systems (Mostafazadeh, 2010). More explicitly, the starting point is the Schrödinger equation

$$i\,\dot{\Psi}(t) = H_{\mathrm{eff}}\Psi(t), \qquad H_{\mathrm{eff}} = \begin{pmatrix} -i\gamma_a & v \\ v & -i\gamma_b \end{pmatrix} \tag{2.71}$$

Here $\gamma_a$ and $\gamma_b$ are real quantities related to the decay rates for the two levels, while the complex parameter $v$ characterizes the radiation–atom interaction. We refer to Ben-Aryeh et al. (2004), Cherbal et al. (2007), and to the references therein, for further details. Here we just want to stress that $H_{\mathrm{eff}} \neq H_{\mathrm{eff}}^{\dagger}$, that the analytical expression of $\Psi(t)$ can be explicitly deduced, and that the resulting time evolution describes a decay of the wave function.

Having these kinds of examples in mind, it seems a natural choice to consider, for instance, the following operator:

$$\hat{h} = \omega_a \hat{n}_a + \omega_b \hat{n}_b + \mu\, a\, b^{\dagger}, \tag{2.72}$$

where $a$, $a^\dagger$, and $\hat{n}_a$ are the operators of a first system $\mathcal{S}_a$, while $b$, $b^\dagger$, and $\hat{n}_b$ are those of a second system, $\mathcal{S}_b$, which in our mind should here play the role of a simplified reservoir. Here we take $\omega_a$, $\omega_b$, and $\mu$ to be real. It is clear that $\hat{h} \neq \hat{h}^\dagger$. To make it self-adjoint, we should add a contribution $\mu\, a^\dagger b$ to its definition. However, it is exactly the lack of this term that makes $\hat{h}$ interesting for us. In fact, as we widely discuss in Chapter 6, the interaction term $\mu\, a\, b^\dagger$ in $\hat{h}$ causes the eigenvalue of the operator $\hat{n}_a$ to decrease (because of the presence of the annihilation operator $a$) and that of $\hat{n}_b$ to increase simultaneously (because of $b^\dagger$). In other words, $\hat{h}$ seems to describe a situation in which $\mathcal{S}_b$ is filled up, while $\mathcal{S}_a$ is emptied out. This seems great, as it would suggest an *effective* procedure to have damping without making use of reservoirs with infinite degrees of freedom. Unfortunately, this very simple approach does not work! This can be explicitly seen if we assume that, also in this case, the time evolution of any fixed operator, $x$, is given by $x(t) = e^{i\hat{h}t} x e^{-i\hat{h}t}$. This is, in a certain sense, a natural choice, but with a serious drawback: $e^{\pm i\hat{h}t}$ are no longer unitary operators. However, rather than a problem, this could be considered as a positive feature of this approach, as we really want to deduce damping! A second technical problem arises because, with this definition of $x(t)$, it is not true in general that $(x(t))^\dagger = (x^\dagger)(t)$, so that $x(t)$ and $x^\dagger(t)$ should be considered as independent dynamical variables. For our simple system we find, for instance,

$$a(t) = a\, e^{-i\omega_a t}, \qquad a^\dagger(t) = a^\dagger\, e^{i\omega_a t} + \frac{\mu}{\omega_b - \omega_a}\, b\left(e^{-i\omega_b t} - 1\right).$$

These operators show, in particular, that $a^\dagger(t)$ is not simply the Hermitian conjugate of $a(t)$. Moreover, it is clear that the mean value of the number operator $\hat{n}_a(t) = a^\dagger(t)\, a(t)$ presents no damping at all. Then we conclude that our simple-minded point of view is simply wrong and suggests, moreover, that the time evolution of the operator $x$ is not simply $x(t) = e^{i\hat{h}t} x e^{-i\hat{h}t}$. This conclusion can also be deduced by another general argument: let us suppose, as in (2.71), that the Schrödinger equation that describes damping is the usual one: $i\dot{\Phi}(t) = \hat{h}\, \Phi(t)$, even if $\hat{h}$ is not self-adjoint. Hence, if $\hat{h}$ does not depend on time, the formal solution of this equation is $\Phi(t) = e^{-i\hat{h}t}\Phi(0)$. As we have pointed out several times in this chapter, what we do observe in a concrete experiment is really the mean value of some relevant observable related to the system. Hence we expect that the time evolution of such a mean value cannot be dependent on the representation chosen. In other words, if $x$ is such an observable, we have

$$\langle\Phi(t), x\,\Phi(t)\rangle = \langle e^{-i\hat{h}t}\Phi(0), x\,e^{-i\hat{h}t}\Phi(0)\rangle >= \langle\Phi(0), e^{i\hat{h}^{\dagger}t}x\,e^{-i\hat{h}t}\Phi(0)\rangle,$$

and this suggests to call $x(t)$, rather than $e^{i\hat{h}t}x e^{-i\hat{h}t}$, the following operator: $x(t) := e^{i\hat{h}^{\dagger}t}x\,e^{-i\hat{h}t}$. It is interesting to notice that this definition cures the anomaly we have seen earlier, that is, the fact that, in general, $(x(t))^{\dagger} \neq \left(x^{\dagger}\right)(t)$. In fact, with this different definition, we easily check that $(x(t))^{\dagger} = \left(x^{\dagger}\right)(t)$, for all possible $x$. On the other hand, in general, it is more difficult to deduce the explicit form of the differential equations describing $x(t)$, as no commutator appears any longer, at least at first sight. However, as we can write $\hat{h} = H_r + iH_i$, with $H_r = \frac{1}{2}\left(\hat{h} + \hat{h}^{\dagger}\right) = H_r^{\dagger}$ and $H_i = \frac{1}{2i}\left(\hat{h} - \hat{h}^{\dagger}\right) = H_i^{\dagger}$, the differential equation of motion for $x(t)$ can be written as

$$\frac{dx(t)}{dt} = e^{i\hat{h}^{\dagger}t}\left(\hat{h}^{\dagger}x - x\hat{h}\right)e^{-i\hat{h}t} = e^{i\hat{h}^{\dagger}t}\left([H_r, x] - i\{H_i, x\}\right)e^{-i\hat{h}t},$$

which involves both a commutator and an anticommutator. This makes the situation more difficult than one might think, and for this reason, we prefer, at least in these notes, to work only with self-adjoint operators. However, few more remarks on possible generalizations along this direction are discussed in Chapter 10.

## 2.8 GREEN'S FUNCTION

In the previous sections, we have considered several tools used in quantum mechanics. Here we give a rather synthetic view to a more mathematical object, the so-called Green's function, and we show how it can be used in the solution of differential equations. For concreteness's sake, here we simply consider the following second-order ordinary differential equation:

$$L\left[y(x)\right] := y''(x) + p(x)y'(x) + q(x)y(x) = f(x). \tag{2.73}$$

The *Green's function* for this equation is that function, $G(x)$, satisfying the equation

$$L\left[G(x)\right] = \delta(x), \tag{2.74}$$

which, of course, has to be thought in the sense of distributions. It follows that, for a sufficiently regular function $f(x)$,

$$y_0(x) := \int_{\mathbb{R}} G(x - s) f(s) ds \qquad (2.75)$$

is a particular solution of the differential equation in Equation 2.73. Indeed we have

$$L[y_0(x)] = \int_{\mathbb{R}} L[G(x - s)] f(s) ds = \int_{\mathbb{R}} \delta(x - s) f(s) ds = f(x).$$

Therefore, finding a particular solution of Equation 2.73 is formally very easy once $G(x)$ is known. What may be hard is to compute $G(x)$ itself. The simplest way to show how, given a differential equation, $G(x)$ can be computed is to consider an explicit example.

**Example 2.2**  *Let us look for a particular solution of the equation* $y'' - 2y = x$.

*It is very easy to find such a solution using only elementary considerations:* $y_0(x) = -\frac{1}{2}x$.

*What we want to do here is to recover this solution computing first the Green's function associated with our differential equation. By definition, $G(x)$ must be such that $L[G(x)] = \delta(x)$, where the operator $L$ is defined as $L = \frac{d^2}{dx^2} - 2$.*

*By introducing the Fourier transform of $G(x)$, $\hat{G}(p)$, we find that*

$$G(x) = \frac{1}{\sqrt{2\pi}} \int_{\mathbb{R}} \hat{G}(p) e^{ipx} dp, \quad \text{so that} \quad L[G(x)]$$

$$= \frac{1}{\sqrt{2\pi}} \int_{\mathbb{R}} \hat{G}(p) \left((ip)^2 - 2\right) e^{ipx} dX.$$

*Then, recalling that the integral expression for the Dirac delta distribution is $\delta(x) = \frac{1}{2\pi} \int_{\mathbb{R}} e^{ipx} dX$, we see that $L[G(x)] = \delta(x)$, if we take $\hat{G}(p) = \frac{1}{\sqrt{2\pi}} \frac{-1}{p^2 + 2}$.*

*Now, to recover $G(x)$, we have to compute the inverse Fourier transform of $\hat{G}(p)$*

$$G(x) = \frac{1}{\sqrt{2\pi}} \int_{\mathbb{R}} \frac{e^{ipx}}{p^2 + 2} \left(-\frac{1}{\sqrt{2\pi}}\right) dp = \frac{-1}{2\pi} \int_{\mathbb{R}} \frac{e^{ipx}}{p^2 + 2} dp.$$

*Using standard techniques in contour integration we find that*

$$
\begin{cases}
G(x > 0) = -\dfrac{1}{2\pi} 2\pi i \, Res \left\{ \dfrac{e^{ipx}}{p^2 + 2}, i\sqrt{2} \right\} = -\dfrac{1}{2\sqrt{2}} e^{-\sqrt{2}x} \\[3mm]
G(x < 0) = -\dfrac{1}{2\pi} (-2\pi i) \, Res \left\{ \dfrac{e^{ipx}}{p^2 + 2}, -i\sqrt{2} \right\} = -\dfrac{1}{2\sqrt{2}} e^{\sqrt{2}x}.
\end{cases}
$$

*Notice that we have also* $G(0) = -\frac{1}{\pi} \int_{\mathbb{R}} \frac{dp}{p^2+2} = -\frac{1}{2\sqrt{2}}$. *Summarizing, we can write* $G(x) = -\frac{1}{2\sqrt{2}} e^{-\sqrt{2}|x|}$. *Hence, Equation 2.75 now becomes*

$$
y_0(x) = -\frac{1}{2\sqrt{2}} \int_{\mathbb{R}} e^{-\sqrt{2}|x-s|} s \, ds = -\frac{1}{2} x,
$$

*as expected.*

Of course, using the Green's function to solve this easy differential equation is not really motivated. Nevertheless, in many problems in quantum mechanics, and in scattering theory in particular, computation of the Green's function is the very first step to get the solution of the problem or, in many interesting cases, to produce a perturbative expansion. We refer to Roman (1965) for many other details and other applications of Green's functions to quantum mechanics. Concerning these notes, the Green's function is used, for instance, in Section 3.4, to compute the solution of the differential equations of motion governing the damped love affair described there.

**PART I**

# SYSTEMS WITH FEW ACTORS

# CHAPTER 3

# LOVE AFFAIRS

## 3.1 INTRODUCTION AND PRELIMINARIES

The first application we consider in this book is, probably, the simplest one as it involves only two or three *actors*. Only in Section 3.4, we consider a significantly large number of people defining, all together, the physical system. Using CCR, we are going to describe what we can call *a love affair*. The main idea is that we can measure the mutual affection of the actors of our model using natural numbers (the higher the number, the stronger the love), and to think that some conserved quantities do exist in the game. One of these integrals of motion is the *global affection*: the total amount of love is not destroyed during the interactions, but only exchanged between the actors. This is true independently of the number of people taking part in the affair. We discuss more about this in the rest of the chapter. It might be worth recalling that sophisticated mathematical tools have been proposed along the years in the analysis of this problem, producing many interesting results that can be found in Strogatz (1988, 1998a,b) and Sproot (2004, 2005), as well as in an extensive monograph (Gottman et al., 2002). For instance, in Sproot (2004), the author considers some simple dynamic models producing coupled ordinary differential

*Quantum Dynamics for Classical Systems: With Applications of the Number Operator*,
First Edition. Fabio Bagarello.
© 2013 John Wiley & Sons, Inc. Published 2013 by John Wiley & Sons, Inc.

equations, which are used for the time evolution of the love (or the hate) in a romantic relationship.

This chapter is organized as follows. In the next section, we consider the first simple model involving two lovers, Alice and Bob, and we analyze the dynamics of their relationship starting from very natural technical and *sentimental* assumptions, and adopting the general tools introduced in the previous chapter. Both a linear and nonlinear model are considered; then, the equations of motion are solved analytically (for the linear model) and numerically (in the nonlinear regime) under suitable (and fairly good) approximations.

In Section 3.3, we consider a model in which Bob has two love relationships at the same time, and again we carry on our dynamical analysis. In addition, in this case, we find an explicit solution for the linear model, while the nonlinear one is analyzed using numerical techniques. In Section 3.4, we go back to Alice and Bob, showing how, *enriching* the model by adding the *environment*, damping effects can be implemented within our framework. We end this chapter with a brief comparison between our approach and those of others.


## 3.2   THE FIRST MODEL

The first model we consider is made by two lovers, Alice and Bob, who mutually *interact* exhibiting a certain interest for each other. Of course, there are several degrees of possible interest, and to a given Bob's interest for Alice (LoA, *level of attraction*), there corresponds a related reaction (i.e., a different LoA) of Alice for Bob. In this section, we show how this mechanism, and the time evolution of Alice–Bob love affair, can be described in terms of creation and annihilation operators.

For that, we introduce $a_1$ and $a_2$, two independent bosonic operators. This means that they obey the commutation rules,

$$[a_i, a_k^\dagger] = a_i\, a_k^\dagger - a_k^\dagger\, a_i = \mathbb{1}\,\delta_{i,k}, \qquad (3.1)$$

where $i, k - 1, 2$, while all the other commutators are trivial: $[a_i, a_k] = [a_i^\dagger, a_k^\dagger] = 0$, for all $i$ and $k$. Further, let $\varphi_0^{(j)}$ be the *vacuum* of $a_j$, $a_j\,\varphi_0^{(j)} = 0$, $j = 1, 2$. By using $\varphi_0^{(j)}$ and the operators $a_j^\dagger$, we may construct, as in Section 2.1, the following vectors:

$$\left.\begin{aligned}
\varphi_{n_j}^{(j)} &:= \frac{1}{\sqrt{n_j!}}\,(a_j^\dagger)^{n_j}\,\varphi_0^{(j)}, \quad \varphi_{n_1,n_2} := \varphi_{n_1}^{(1)} \otimes \\
\varphi_{n_2}^{(2)} &= \frac{1}{\sqrt{n_1!\,n_2!}}\,(a_1^\dagger)^{n_1}\,(a_2^\dagger)^{n_2}\,\varphi_{0,0},
\end{aligned}\right\} \qquad (3.2)$$

where $\varphi_{0,0} = \varphi_0^{(1)} \otimes \varphi_0^{(2)}$, $n_j = 0, 1, 2, \ldots$, and $j = 1, 2$. Let us also introduce the number operators $\hat{n}_j = a_j^\dagger a_j$, $j = 1, 2$, and $\hat{N} = \hat{n}_1 + \hat{n}_2$. Hence, as discussed in Section 2.1, we get

$$\begin{cases} \hat{n}_j \varphi_{n_1,n_2} = n_j \varphi_{n_1,n_2}, \\ \hat{n}_j [a_j \varphi_{n_1,n_2}] = (n_j - 1)[a_j \varphi_{n_1,n_2}], \\ \hat{n}_j [a_j^\dagger \varphi_{n_1,n_2}] = (n_j + 1)[a_j^\dagger \varphi_{n_1,n_2}]. \end{cases} \tag{3.3}$$

where $j = 1, 2$. Then, we also have $\hat{N}\varphi_{n_1,n_2} = (n_1 + n_2)\varphi_{n_1,n_2}$. As usual, the Hilbert space, $\mathcal{H}$, in which the operators live, is obtained by taking the closure of the linear span of all these vectors, for $n_j \geq 0$, $j = 1, 2$. In other words, the set $\mathcal{F}_\varphi := \{\varphi_{n_1,n_2}, n_1, n_2 \geq 0\}$ is an o.n. basis of $\mathcal{H}$, whose dimension is, as a consequence, infinite (but numerable). A state over the system is a normalized linear functional $\omega_{n_1,n_2}$ labeled by two *quantum numbers* $n_1$ and $n_2$ such that $\omega_{n_1,n_2}(x) = \langle \varphi_{n_1,n_2}, x\, \varphi_{n_1,n_2}\rangle$, where $\langle .,.\rangle$ is the scalar product in $\mathcal{H}$ and $x$ is any given operator on $\mathcal{H}$.

In this chapter, we associate the (integer) eigenvalue $n_1$ of $\hat{n}_1$ to the LoA that Bob experiences for Alice: the higher the value of $n_1$, the more Bob desires Alice. For instance, if $n_1 = 0$, Bob just does not care about Alice. We use $n_2$, the eigenvalue of $\hat{n}_2$, to label the attraction of Alice for Bob. A well-known (even if surely oversimplified) law of attraction stated in our language simply says that if $n_1$ increases, $n_2$ decreases and vice versa.[1] This suggests using the following self-adjoint operator to describe the dynamics of the relationship, see **Rule 1** below:

$$H = \lambda \left( a_1^{M_1} a_2^{\dagger M_2} + a_2^{M_2} a_1^{\dagger M_1} \right). \tag{3.4}$$

Here, $M_1$ and $M_2$ give a measure of the kind of mutual reaction between Bob and Alice: if $M_1$ is large compared to $M_2$, Bob will change his status much faster than Alice. The opposite change is expected for $M_2$ much larger than $M_1$, while, for $M_1$ close to $M_2$, Alice and Bob will react essentially with the same speed. These claims will be justified in a moment. In the meantime, let us consider a first consequence: it is enough to introduce a single power $M$, rather than $M_1$ and $M_2$, with $M$ playing the

---

[1]This law has inspired many (not only) Italian love songs over the years!

role of a *relative behavior*. For this reason, we simplify the Hamiltonian (Eq. 3.4) replacing it with

$$H = \lambda \left( a_1^M a_2^\dagger + a_2 a_1^{\dagger M} \right),$$    (3.5)

where $\lambda$ is the interaction parameter (which, as will be clear in the following, could also be seen as a time scaling parameter). The physical meaning of $H$ can be deduced considering, for instance, the action of, say, $a_1^M a_2^\dagger$ on the vector describing the system at time $t = 0$, $\varphi_{n_1,n_2}$. This means that, at $t = 0$, Bob is in the state $n_1$, that is, $n_1$ is Bob's LoA, whereas Alice is in the state $n_2$. For instance, if $n_1 > n_2$, then Bob loves Alice more than Alice loves Bob. However, because of the definition of $\varphi_{n_1,n_2}$, $a_1^M a_2^\dagger \varphi_{n_1,n_2}$, which is different from zero only if $M \le n_1$, is proportional to $\varphi_{n_1-M,n_2+1}$. Hence, after the action of $a_1^M a_2^\dagger$, Bob's interest for Alice decreases by $M$ units, whereas Alice's interest for Bob increases by 1 unit. Of course, the Hamiltonian (Eq. 3.5) also describes the opposite behavior. Indeed, because of the presence of $a_2 a_1^{\dagger M}$ in $H$, if $n_2 \ge 1$, we see that $a_2 a_1^{\dagger M} \varphi_{n_1,n_2}$ is proportional to $\varphi_{n_1+M,n_2-1}$. Therefore, Bob's interest increases (by $M$ units) whereas Alice loses interest in Bob. It is not hard to check that $I(t) := \hat{n}_1(t) + M \hat{n}_2(t)$ is a constant of motion: $I(t) = I(0) = \hat{n}_1(0) + M \hat{n}_2(0)$, for all $t \in \mathbb{R}$. This is a consequence of the meaning of $H$ or, more explicitly, of the following commutation result: $[H, I] = 0$. Therefore, during the time evolution, a certain *global attraction* is preserved and can only be exchanged between Alice and Bob. Notice that this reproduces our original point of view on how a (simplified) love relation essentially works. In particular, if $M = 1$, the constant of motion becomes $I(t) := \hat{n}_1(t) + \hat{n}_2(t)$: the affection lost by Alice is gained by Bob, and vice versa.

Now, we have all the ingredients to derive the equations of motion for our model. These are found by assuming, as we have widely discussed in Chapter 2, the same Heisenberg-like dynamics that works perfectly for quantum systems, and which is somehow also a natural choice in the present operatorial settings. Of course, this is a very strong assumption and should be checked *a posteriori*, finding the dynamical behavior deduced in this way and showing that this gives reasonable results. As the Hamiltonian $H$ in Equation 3.5 does not depend explicitly on time, the time evolution $X(t)$ of a given observable $X$ of the system is given by $X(t) = e^{iHt} X e^{-iHt}$, see Equation 2.1, or, equivalently, by the solution of the differential equation $\dot{X}(t) = i[H, X(t)]$, with $X(0) = X$, see Equation 2.2.

The equations of motion for the number operators $\hat{n}_1(t)$ and $\hat{n}_2(t)$, which are needed to deduce the *rules of the attraction*, are the following:

$$
\begin{cases}
\dfrac{\mathrm{d}}{\mathrm{d}t}\,\hat{n}_1(t) = i\,\lambda M\left(a_2^\dagger(t)(a_1(t))^M - a_2(t)(a_1(t)^\dagger)^M\right), \\[2ex]
\dfrac{\mathrm{d}}{\mathrm{d}t}\,\hat{n}_2(t) = -i\,\lambda\left(a_2^\dagger(t)(a_1(t))^M - a_2(t)(a_1(t)^\dagger)^M\right).
\end{cases}
\tag{3.6}
$$

By using this system, it is straightforward to check explicitly that $I(t)$ does not depend on time, as stated earlier: in fact, recalling the definition of $I(t)$, we have $\dot{I}(t) = \frac{\mathrm{d}}{\mathrm{d}t}\,(\hat{n}_1(t) + M\,\hat{n}_2(t)) = 0$. However, the differential equations in 3.6 are not closed, so that it may be more convenient to replace them by the differential system for the annihilation operators $a_1(t)$ and $a_2(t)$:

$$
\begin{cases}
\dfrac{\mathrm{d}}{\mathrm{d}t}\,a_1(t) = -i\,\lambda M\,a_2(t)(a_1(t)^\dagger)^{M-1} \\[2ex]
\dfrac{\mathrm{d}}{\mathrm{d}t}\,a_2(t) = -i\,\lambda\,(a_1(t))^M.
\end{cases}
\tag{3.7}
$$

Then, we may use the solutions of this system to construct the number operators $\hat{n}_j(t) = a_j^\dagger(t)\,a_j(t)$, $j = 1, 2$. Equation 3.7, together with its adjoints, now produces a closed system. Of course, there exists a simple situation for which Equation 3.7 can be quite easily solved analytically: $M = 1$. In this case, which corresponds to the assumption that Alice and Bob react *with the same speed*, Equation 3.7 is already closed and the solution is easily found:

$$
a_1(t) = a_1 \cos(\lambda t) - i a_2 \sin(\lambda t), \qquad a_2(t) = a_2 \cos(\lambda t) - i a_1 \sin(\lambda t).
\tag{3.8}
$$

Now, if we assume that at $t = 0$ Bob and Alice are, respectively, in the $n_1$th and $n_2$th LoAs, this is equivalent to say that the system $\mathcal{S}$ at $t = 0$ is described by the vector $\varphi_{n_1,n_2}$ so that the corresponding state is $\omega_{n_1,n_2}(\cdot) = \langle \varphi_{n_1,n_2}, \cdot\, \varphi_{n_1,n_2}\rangle$. This is our way to introduce in the game the initial conditions describing the $t = 0$ status of the two lovers. Therefore, calling $n_j(t) := \omega_{n_1,n_2}(\hat{n}_j(t))$, $j = 1, 2$, we find that

$$
n_1(t) = n_1 \cos^2(\lambda t) + n_2 \sin^2(\lambda t), \qquad n_2(t) = n_2 \cos^2(\lambda t) + n_1 \sin^2(\lambda t),
\tag{3.9}
$$

so that, in particular, $\omega_{n_1,n_2}(I(t)) = n_1 + n_2$, as expected. Incidentally, these equations imply that $n_1(0) = n_1$ and $n_2(0) = n_2$, which are exactly our initial conditions. The conclusion is quite simple and close to our view of how the law of attraction works: the infatuations of Alice and Bob oscillate in such a way that whenever Bob's LoA increases, Alice's LoA decreases, and vice versa, with a period that is directly related to the value of the interaction parameter $\lambda$. In particular, as it is natural, setting $\lambda = 0$ in Equation 3.9 implies that both Alice and Bob stay in their initial LoAs. It is also clear now that $\lambda$ behaves as a time scaling parameter. As it is clear, the solution in Equation 3.9 is the first *a posteriori* justification of our approach: the law of attraction between Alice and Bob is all contained in a single operator, the Hamiltonian of the model, whose explicit expression can be easily deduced using rather general arguments. The related dynamics, that is, Equation 3.9, is exactly the one we expected to find. Hence, the use of Heisenberg equations of motion seems to be justified, at least for this simple model. This is in agreement with what was preliminarily stated in Section 1.2. Similar results were also found for the linear model in Sproot (2004) under suitable assumptions on the parameters.

Much harder is the situation when $M > 1$. In this case, we need to consider the adjoint of Equation 3.7 to close the system and, nevertheless, an exact analytical solution cannot be obtained. However, it is possible to generate a numerical scheme to find solutions of our problem, and this is the content of the next subsection. It might be worth stressing that the difficulty of solving analytically the differential equations for the nonlinear system has nothing to do with the operatorial approach we are adopting here: nonlinear differential equations are usually very difficult to be solved, even when the unknown is a *simple* function of time.

### 3.2.1 Numerical Results for $M > 1$

We start remarking that, as stated earlier, the Hilbert space $\mathcal{H}$ of our theory is infinite-dimensional. This means that both Bob and Alice may experience, in principle, infinite different LoAs. This makes the situation rather hard from a computational point of view and, furthermore, appears as an unnecessary complication. As a matter of fact, it is enough to assume that Bob (respectively, Alice) may pass through $L_1$ (respectively, $L_2$) different LoAs ($L_1$ and $L_2$ fixed positive integers), which efficiently describe their mutual attraction. This can be easily understood recalling the existence of the integral of motion $I(t)$. If Alice's and Bob's LoAs are, respectively, $n_2$ and $n_1$ at $t = 0$, then the mean value of $I(0)$, $\omega_{n_1,n_2}(I(0))$, is $n_1 + M n_2$. Hence, $L_1$ and $L_2$ are constrained by the following inequalities: $L_1 \leq n_1 + M n_2$, and $L_2 \leq \frac{1}{M}(n_1 + M n_2)$. Choosing $L_1$ and $L_2$ larger than the upper bounds given by these inequalities is not needed because these levels cannot be reached; otherwise, this would violate the existence of the integral of motion $I(t)$. Hence, we deduce that our *effective Hilbert space*, $\mathcal{H}_{\text{eff}}$, is finite-dimensional and is generated by the o.n. basis

$$\mathcal{F} = \{\varphi_{\underline{n}} := \varphi_{n_1, n_2}, \quad n_j = 0, 1, \ldots, L_j, \ j = 1, 2\} = \{\varphi_{\underline{n}}, \quad \underline{n} \in \mathcal{K}\}, \tag{3.10}$$

with obvious notation. Hence, the dimension of $\mathcal{H}_{\text{eff}}$, $dim(\mathcal{H}_{\text{eff}}) = (L_1 + 1)(L_2 + 1)$, is exactly the cardinality of $\mathcal{K}$.

Calling $\mathbb{1}_{\text{eff}}$ the identity operator over $\mathcal{H}_{\text{eff}}$, the closure relation for $\mathcal{F}$ looks like

$$\sum_{\underline{n} \in K} |\varphi_{\underline{n}}><\varphi_{\underline{n}}| = \mathbb{1}_{\text{eff}}.$$

In this formula, we are using the so-called Dirac *bra-ket notation*. Let us see in some detail what is the meaning of this formula. First of all, we notice that its left-hand side is the sum of operators $|\varphi_n><\varphi_{\underline{n}}|$. These are defined in the following way: let $\Phi, \eta, \Psi \in \mathcal{H}_{\text{eff}}$. Then

$$(|\Phi><\Psi|)\eta := \langle \Psi, \eta \rangle \Phi.$$

This means that $|\Phi><\Psi|$ projects any vector of $\mathcal{H}_{\text{eff}}$ into a one-dimensional subspace generated by $\Phi$. With the above definitions, we can easily check that $(|\varphi_n><\varphi_{\underline{n}}|)^2 = |\varphi_n><\varphi_{\underline{n}}|$, and that $|\varphi_n><\varphi_{\underline{n}}|$ is self-adjoint. For instance, for each $f \in \mathcal{H}_{\text{eff}}$,

$$\left(|\varphi_{\underline{n}}><\varphi_{\underline{n}}|\right)^2 f = \left(|\varphi_{\underline{n}}><\varphi_{\underline{n}}|\right)\left(\langle \varphi_{\underline{n}}, f \rangle \varphi_{\underline{n}}\right)$$

$$= \langle \varphi_{\underline{n}}, f \rangle \langle \varphi_{\underline{n}}, \varphi_{\underline{n}} \rangle \varphi_{\underline{n}} = \left(|\varphi_{\underline{n}}><\varphi_{\underline{n}}|\right) f,$$

where we have used the fact that each $\varphi_n$ is normalized. Then, for each $\underline{n}$, $|\varphi_n><\varphi_{\underline{n}}|$ is an *orthogonal projection operator*.

It is a standard exercise in quantum mechanics to check that

$$\begin{cases} (a_1)_{\underline{k}, \underline{k}'} := <\varphi_{\underline{k}}, a_1 \varphi_{\underline{k}'}> = \sqrt{k_1'} \, \delta_{k_1, k_1'-1} \, \delta_{k_2, k_2'}, \\ (a_2)_{\underline{k}, \underline{k}'} := <\varphi_{\underline{k}}, a_2 \varphi_{\underline{k}'}> = \sqrt{k_2'} \, \delta_{k_2, k_2'-1} \, \delta_{k_1, k_1'}, \end{cases} \tag{3.11}$$

so that, by using the resolution of the identity in $\mathcal{H}_{\text{eff}}$, we can produce a *projected* expression of the operators $a_j$ as follows:

$$
\begin{aligned}
a_1 \to A_1 &:= \sum_{k,k' \in K} <\varphi_{\underline{k}}, a_1 \varphi_{\underline{k}'}> |\varphi_{\underline{k}}> <\varphi_{\underline{k}'}| \\
&= \sum_{k' \in K} \sqrt{k'_1} |\varphi_{k'_1 - 1, k'_2}> <\varphi_{k'_1 k'_2}|, \\
a_2 \to A_2 &:= \sum_{k,k' \in K} <\varphi_{\underline{k}}, a_2 \varphi_{\underline{k}'}> |\varphi_{\underline{k}}> <\varphi_{\underline{k}'}| \\
&= \sum_{k' \in K} \sqrt{k'_2} |\varphi_{k'_1, k'_2 - 1}> <\varphi_{k'_1 k'_2}|.
\end{aligned}
\tag{3.12}
$$

In other words, whereas the $a_j$s act on $\mathcal{H}$, the related matrices $A_j$s act on the finite-dimensional Hilbert space $\mathcal{H}_{\text{eff}}$. This implies, in particular, that whereas the $a_j(t)$s are unbounded operators (which could be represented by infinite matrices), the $A_j(t)$s are $(L_1 + 1)(L_2 + 1) \times (L_1 + 1)(L_2 + 1)$ matrices whose elements, for $t = 0$, can be deduced by Equation 3.12. Thus, these are really bounded operators and, as we see, they all share the following property: for all choices of $L_j$, a certain power $R_j$ can be found such that $A_j^{R_j} = 0$ and $(A_j^\dagger)^{R_j} = 0$. A similar exponent does not exist, in general, for the original operators $a_j$ and $a_j^\dagger$. This fact is related to the approximation that produces $\mathcal{H}_{\text{eff}}$ out of the original $\mathcal{H}$, and can be easily understood in the following way: as $\mathcal{H}_{\text{eff}}$ has only $L_1 + 1$ possible levels for Bob's LoA, if he is already in the $L_1$th level, acting once more with $A_1^\dagger$ on the vector of the system, this simply destroys the vector. In other words, Bob cannot move to level $L_1 + 1$, as the related vector does not belong to $\mathcal{H}_{\text{eff}}$. Now, Equation 3.7 can be written in a formally identical way as

$$
\begin{cases}
\dot{A}_1(t) = -i\,\lambda M\, A_2(t)(A_1(t)^\dagger)^{M-1}, \\
\dot{A}_2(t) = -i\,\lambda\,(A_1(t))^M,
\end{cases}
\tag{3.13}
$$

where $A_1(t)$ and $A_2(t)$ are bounded matrices rather than unbounded operators. It should also be stressed that the change $a_j \to A_j$ does not destroy the existence of an integral of motion, which, not surprisingly, turns out to be $I_{\text{eff}}(t) = \mathcal{N}_1(t) + M\mathcal{N}_2(t)$, where $\mathcal{N}_j(t) = A_j(t)^\dagger A_j(t)$.

The numerical scheme is quite simple: we just have to fix the dimensionality of $\mathcal{H}_{\text{eff}}$, that is, the values of $L_1$ and $L_2$ (i.e., the accessible number of LoAs), the value of $M$ in Equation 3.5, and the matrices $A_j(0)$. Then, by choosing a reliable scheme for integrating numerically a set of ordinary differential equations, we may construct a solution for Equation 3.13 in a prescribed time interval. From now on, we take $L_1 = L_2 = K$ ($K$ suitable positive integer), whereupon $dim(\mathcal{H}_{\text{eff}}) = (K+1)^2$, and we consider the o.n. basis of $\mathcal{H}_{\text{eff}}$ in the following order (the order is important to fix the form of the matrices!):

$$\mathcal{F} = \left\{ \varphi_{00}, \varphi_{01}, \ldots, \varphi_{0K}, \varphi_{10}, \varphi_{11}, \ldots, \varphi_{1K}, \ldots, \varphi_{K0}, \varphi_{K1}, \ldots, \varphi_{KK} \right\}.$$

Then we get

$$A_1(0) = \begin{pmatrix} \mathbf{0}_K & \mathbf{0}_K & \cdots & \cdots & \mathbf{0}_K & \mathbf{0}_K \\ \mathbb{1}_K & \mathbf{0}_K & \cdots & \cdots & \mathbf{0}_K & \mathbf{0}_K \\ \mathbf{0}_K & \sqrt{2}\,\mathbb{1}_K & \mathbf{0}_K & \cdots & \mathbf{0}_K & \mathbf{0}_K \\ \mathbf{0}_K & \mathbf{0}_K & \sqrt{3}\,\mathbb{1}_K & \cdots & \mathbf{0}_K & \mathbf{0}_K \\ \cdots & \cdots & \cdots & \cdots & \cdots & \cdots \\ \cdots & \cdots & \cdots & \cdots & \cdots & \cdots \\ \mathbf{0}_K & \mathbf{0}_K & \cdots & \cdots & \sqrt{K-1}\,\mathbb{1}_K & \mathbf{0}_K \end{pmatrix},$$

$$A_2(0) = \begin{pmatrix} \underline{x} & \mathbf{0}_K & \cdots & \cdots & \mathbf{0}_K \\ \mathbf{0}_K & \underline{x} & \cdots & \cdots & \mathbf{0}_K \\ \cdots & \cdots & \cdots & \cdots & \cdots \\ \cdots & \cdots & \cdots & \cdots & \cdots \\ \mathbf{0}_K & \mathbf{0}_K & \cdots & \cdots & \underline{x} \end{pmatrix},$$

where $\mathbf{0}_K$ and $\mathbb{1}_K$ are the null and the identity matrices of order $K$, respectively, whereas

$$\underline{x} = \begin{pmatrix} 0 & 0 & 0 & \cdots & 0 \\ 1 & 0 & 0 & \cdots & 0 \\ 0 & \sqrt{2} & 0 & \cdots & 0 \\ \cdots & \cdots & \cdots & \cdots & \cdots \\ 0 & 0 & \cdots & \sqrt{K-1} & 0 \end{pmatrix}.$$

It is easy to verify that $A_1(0)^{K-1} \neq \mathbf{0}_K$ and $A_2(0)^{K-1} \neq \mathbf{0}_K$, whereas $A_1(0)^K = A_2(0)^K = \mathbf{0}_K$. Similar conclusions were deduced for the *effective creation operators* $A_j^\dagger(0)$, $j = 1, 2$. As already discussed, this simply means that because of our approximation $\mathcal{H} \to \mathcal{H}_{\text{eff}}$, there are only $K+1$ different levels in our Hilbert space; then, if we act too many times on a certain state, the only effect we get is just to annihilate that state. Stated differently, we cannot move Bob or Alice to a $-1$ or to a $K+1$ LoA by acting with $A_j$ or $A_j^\dagger$ as these states do not exist in $\mathcal{H}_{\text{eff}}$! Notice that a similar problem also exists for the operators $a_j$ (but not for $a_j^\dagger$): if we act *too many times* with the annihilation operator $a_j$ on a given vector $\varphi_{n_1,n_2}$, we end up destroying that vector.[2]

---

[2]This happens when $a_1$ acts on $\varphi_{n_1,n_2}$ $n_1 + 1$ times, or when $a_2$ acts on $\varphi_{n_1,n_2}$ $n_2 + 1$ times.

Suppose now that, for $t = 0$, the system is in the state $\omega_{\underline{n}}$ introduced before, where $\underline{n} = (n_1, n_2)$. As we have done explicitly for $M = 1$, if we want to know how Bob's LoA varies with time, we have to compute

$$n_1(t) := \, < \varphi_{\underline{n}}, \mathcal{N}_1(t)\varphi_{\underline{n}} > \, = \|A_1(t)\varphi_{\underline{n}}\|^2 = \sum_{\underline{k} \in K} |(A_1(t))_{\underline{k},\underline{n}}(t)|^2,$$

where $<,>$ is the scalar product in $\mathcal{H}_{\text{eff}}$, and $\|\,\|$ its related norm, in terms of the elements of the $A_1(t)$. Analogously, to compute how Alice's LoA varies with time, we need to compute

$$n_2(t) := \, < \varphi_{\underline{n}}, \mathcal{N}_2(t)\varphi_{\underline{n}} > \, = \|A_2(t)\varphi_{\underline{n}}\|^2 = \sum_{\underline{k} \in K} |(A_2(t))_{\underline{k},\underline{n}}(t)|^2.$$

Equation 3.13 is numerically integrated for different choices of $K$ by taking $M = 2$ and $\lambda = 0.1$. In particular, we have used a variable-order Adams–Bashforth–Moulton PECE solver (Shampine et al., 1975) as implemented in MATLAB®'s ode113 routine.

In Figure 3.1, we plot the time evolutions of Alice's and Bob's LoAs in the case $K = 3$ with equal initial conditions $n_1(0) = n_2(0) = 1$. Two clear oscillations in opposite phase can be observed: Alice and Bob react simultaneously but *in different directions*. This is in agreement with our naive point of view of Alice–Bob's love relationship, and confirms that the Heisenberg equations of motion can reasonably be used to describe the dynamics of this classical system, also in this nonlinear (and possibly more interesting) case. The horizontal dotted line on the top in Figure 3.1 (and also in Figures 3.2 and 3.3) represents the integral of motion $I_{\text{eff}}(t)$; in some sense, it provides a check of the quality of the numerical solution because, as predicted by the general arguments discussed before, $I_{\text{eff}}(t)$ must stay constant in time.

It is now necessary to spend some words on our plots and on how they should be interpreted. The reason is the following: from Figure 3.1, we see that Bob's LoA apparently takes all real values between 1 and 3 in the time interval $[0, 10]$. So, in particular, at a certain time, the value of $n_1$ is 1.5. This is not what one could imagine as $n_1(t)$ and $n_2(t)$ were assumed to take only integer values, being the eigenvalues of two number operators. For this reason, the correct interpretation of the curves in Figure 3.1 is simply that *they interpolate between the integer values of the eigenvalues allowed by our framework*. It is obvious that the other plots produced in this book should be considered from the same point of view.

In Figure 3.2, the evolution of Alice's and Bob's LoAs with initial conditions $(2, 2)$ and for different dimensions of $\mathcal{H}_{\text{eff}}$ is plotted. As it

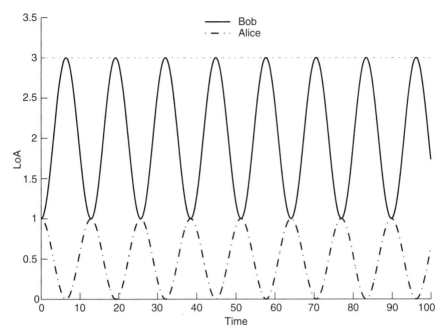

**Figure 3.1**  $K = 3$, $M = 2$: Alice's and Bob's LoAs versus time with initial condition $(1, 1)$. A periodic behavior is observed.

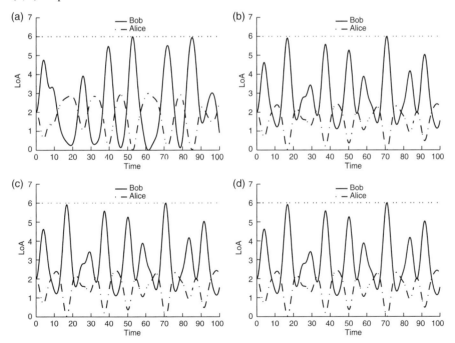

**Figure 3.2**  Alice's and Bob's LoAs versus time with initial conditions $(2, 2)$ and $M = 2$: $K = 3$ (a), $K = 6$ (b), $K = 7$ (c), $K = 8$ (d); $K = 6$ is already a good approximation.

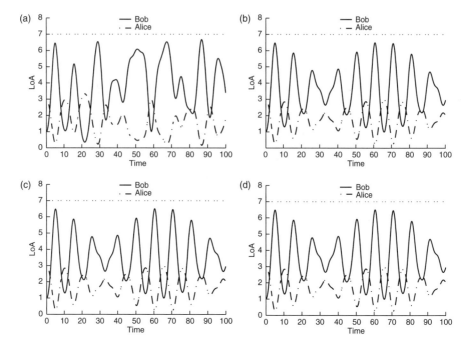

**Figure 3.3**   Alice's and Bob's LoAs versus time with initial condition $(1, 3)$ and $M = 2$: $K = 5$ (a), $K = 6$ (b), $K = 7$ (c), $K = 8$ (d); again, $K = 6$ is already a good approximation.

can be observed, when $K = 3$ (Figure 3.2a), the values of LoAs go beyond the maximum admissible value $(K)$; this suggests to improve the approximation, that is, to increase the dimension of $\mathcal{H}_{\text{eff}}$. By taking $K = 6$ (Figure 3.2b), we find that Alice's and Bob's LoAs assume values within the bounds. Moreover, by performing the numerical integration with higher values of $K$ ($K = 7$ and $K = 8$, Figure 3.2c,d), the same dynamics as when $K = 6$ is recovered. We also see that, as in Figure 3.1, $n_1(t)$ increases when $n_2(t)$ decreases, and vice versa, as we expected because of the existence of $I_{\text{eff}}(t)$: our general *law of attraction* is respected for this nonlinear interaction also.

Essentially the same main conclusions are deduced by Figure 3.3, where the initial condition $(1, 3)$ and different values of $K$ ($K = 5, 6, 7, 8$) are considered. The choice $K = 5$ turns out to be a poor approximation; in fact, the values of LoAs cannot be described in the related $\mathcal{H}_{\text{eff}}$. On the contrary, already for $K = 6$, the associated effective Hilbert space $\mathcal{H}_{\text{eff}}$ turns out to be a fairly good substitute of $\mathcal{H}$ as the values of LoAs do not exceed the due limits during the time evolution. Furthermore, the dynamics is *stable* for increasing values of $K$. This is the same stability that was already observed in Figure 3.2.

Let us summarize the situation: as a consequence of our approximation, $\mathcal{H} \to \mathcal{H}_{\text{eff}}$, we need to replace $I(t)$ with $I_{\text{eff}}(t) = \mathcal{N}_1(t) + M\,\mathcal{N}_2(t)$. As already noticed, because of Equation 3.13, $\dot{I}_{\text{eff}}(t) = 0$, that is, $I_{\text{eff}}(t)$ is a constant of motion for the approximated model. Hence, it happens that if $\mathcal{N}_2(t)$ decreases during its time evolution, $\mathcal{N}_1(t)$ must increase for $I_{\text{eff}}(t)$ to stay constant. In this way, for some value of $t$, it may happen that $\mathcal{N}_1(t) > K$ (see Figure 3.2 for $K = 3$ and Figure 3.3 for $K = 5$). This problem is easily cured by fixing higher values of $K$. The numerical evidence of this fact is given by Figures 3.2 and 3.3, where we have shown that the choice $K = 6$ allows us to capture the right dynamics, and that this dynamics remains unchanged even when $K = 7$ and $K = 8$. We see that, as already stated, the dimension of $\mathcal{H}_{\text{eff}}$ can be fixed *a priori* looking at the value of the integral of motion of the system for $t = 0$: therefore, $dim(\mathcal{H}_{\text{eff}})$ is not a free choice. In particular, it must be greater than or equal to $I_{\text{eff}}(0)$. Last, but not least, our numerical tests allow us to conjecture that, increasing the dimension of $\mathcal{H}_{\text{eff}}$, the resulting dynamics is not affected: hence a sort of *optimal effective Hilbert space* seems to exist.

Of course, this is a consequence of the numerical approach adopted here, and has nothing to do with our general framework. This is clear, for instance, considering the solution of the linear situation where no approximation is needed and no problem with the dimensionality of $\mathcal{H}_{\text{eff}}$ arises (as there was no need to introduce $\mathcal{H}_{\text{eff}}$!). Once again, it could be worth stressing that these kinds of problems, quite often, if not always, appear when dealing with nonlinear interactions.

In this nonlinear case, the numerical results seem to show that a periodic motion is not recovered for all initial conditions, contrary to what was analytically proved in the case $M = 1$. In Figures 3.2 and 3.3, rather than a simple periodic time evolution (similar to that displayed in Figure 3.1), a quasiperiodic behavior seems to emerge, as the plots of the power spectra of the time series representing the numerical solutions (see Figure 3.4c,d) also suggest. However, what emerges from our numerical tests is that once the dimension of $\mathcal{H}_{\text{eff}}$ is sufficient to capture the dynamics, only periodic or quasiperiodic solutions are obtained. This does not exclude that, for initial conditions requiring dimensions of the Hilbert space higher than those considered here, a richer dynamics (say, chaotic) might arise.

## 3.3  A LOVE TRIANGLE

In this section, we generalize our previous model by inserting a third ingredient. Just to fix the ideas, we assume that this third ingredient, Carla,

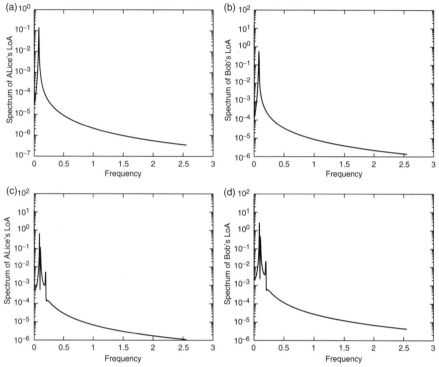

**Figure 3.4** Power spectra versus frequency of numerical solutions for Alice's LoA (a,c) and Bob's LoA (b,d) for initial conditions $(1, 1)$ (a,b: periodic dynamics) and $(1, 3)$ (c,d: quasiperiodic dynamics).

is Bob's lover, and we use the same technique to describe the interactions among the three. Our vision of this love affair can be summarized as follows:

1. Bob can interact with both Alice and Carla, but Alice does not suspect of Carla's role in Bob's life, and vice versa.
2. If Bob's LoA for Alice increases, then Alice's LoA for Bob decreases, and vice versa.
3. Analogously, if Bob's LoA for Carla increases, then Carla's LoA for Bob decreases, and vice versa.
4. If Bob's LoA for Alice increases, then his LoA for Carla decreases (not necessarily by the same amount), and vice versa.

In order to simplify the computations, we assume for the moment that the *action and the reaction* of the lovers have the same strength. Recalling our previous considerations on the meaning of the exponent

$M$ in definition (Eq. 3.5), this suggests to fix, at first, the *nonlinearity parameter* $M$ to be equal to $1$.[3] The Hamiltonian of the system is a simple generalization of that in Equation 3.5 with $M = 1$:

$$H = \lambda_{12} \left( a_{12}^\dagger a_2 + a_{12} a_2^\dagger \right) + \lambda_{13} \left( a_{13}^\dagger a_3 + a_{13} a_3^\dagger \right)$$

$$+ \lambda_1 \left( a_{12}^\dagger a_{13} + a_{12} a_{13}^\dagger \right). \tag{3.14}$$

Here the indices 1, 2, and 3 stand for Bob, Alice, and Carla, respectively, the $\lambda_\alpha$s are coefficients measuring the relative interaction strengths, and are taken real so to have $H = H^\dagger$, and the different $a_\alpha$, $\alpha = 12, 13, 2, 3$, are bosonic operators. They satisfy the usual CCR $[a_\alpha, a_\beta^\dagger] = \delta_{\alpha,\beta} \mathbb{1}$, whereas all the other commutators are zero. The three terms in $H$ reflect, respectively, points 2, 3, and 4 of the above list. As we have discussed in Chapter 1, one of the positive aspects of the general approach discussed in this book is that the Hamiltonian of the system, which is the only ingredient needed in the deduction of the dynamics, can be written in a quite simple way starting from very general assumptions, such as the ones listed earlier. More explicitly, the recipe we have adopted here to write down $H$, and which will be used everywhere, sometimes in a slightly more complicated version, can be summarized as follows:

**Rule 1:** *Suppose that our system, $\mathcal{S}$, consists of two (or more) actors, $\Theta_1$ and $\Theta_2$, interacting and exchanging something (money, shares, love, ... ). We associate two annihilation operators, $a_1$ and $a_2$, and their adjoints, respectively, to $\Theta_1$ and $\Theta_2$. Then the exchange is modeled adding to the Hamiltonian of $\mathcal{S}$ a term $a_1^\dagger a_2 + a_2^\dagger a_1$. If, for some reason, the model should be nonlinear, then this contribution must be replaced by $a_1^{\dagger^M} a_2 + a_2^\dagger a_1^M$, $M = 1, 2, 3, \ldots$ being a measure of the nonlinearity.*

We use this general recipe, together with other rules that will be introduced shortly, several times along these notes. Moreover, in the second part of this book, this first general idea is extended to consider also the possibility of exchanging not just a single quantity (here is the *love* that goes back and forth between Alice, Bob, and Carla). For instance, in Part II, the actors are the traders of our SSM, and they will exchange money

---

[3]Notice that, in this love triangle, more than one such parameter should be used. This is exactly what we will do in Section 3.3.1.

and shares at the same time. We will return to this rule in Chapter 6, which is entirely devoted to the construction of the Hamiltonian of a given system.

Going back to our model, we also introduce the number operators $\hat{n}_{12} = a_{12}^\dagger a_{12}$, describing Bob's LoA for Alice; $\hat{n}_{13} = a_{13}^\dagger a_{13}$, describing Bob's LoA for Carla; $\hat{n}_2 = a_2^\dagger a_2$, describing Alice's LoA for Bob; and $\hat{n}_3 = a_3^\dagger a_3$, describing Carla's LoA for Bob. If we define $J := \hat{n}_{12} + \hat{n}_{13} + \hat{n}_2 + \hat{n}_3$, which represents the global LoA of the triangle, this is a conserved quantity: $J(t) = J(0)$, as $[H, J] = 0$. It is also possible to check that $[H, \hat{n}_{12} + \hat{n}_{13}] \neq 0$ so that the total Bob's LoA is not conserved during the time evolution.

The equations of motion for the variables $a_\alpha$s can be deduced as usual, and we find:

$$\begin{cases} i\,\dot{a}_{12}(t) = \lambda_{12}\,a_2(t) + \lambda_1\,a_{13}(t), \\ i\,\dot{a}_{13}(t) = \lambda_{13}\,a_3(t) + \lambda_1\,a_{12}(t), \\ i\,\dot{a}_2(t) = \lambda_{12}\,a_{12}(t), \\ i\,\dot{a}_3(t) = \lambda_{13}\,a_{13}(t). \end{cases} \tag{3.15}$$

This linear system can be explicitly solved, and the solution can be written as

$$A(t) = U^{-1} \exp\left(-i\Lambda_d\,t\right) U\,A(0), \tag{3.16}$$

where $A$ is the (column) vector with components $(a_{12}, a_{13}, a_2, a_3)$, $\Lambda_d$ is the diagonal matrix of the eigenvalues of the matrix

$$\Lambda = \begin{pmatrix} 0 & \lambda_1 & \lambda_{12} & 0 \\ \lambda_1 & 0 & 0 & \lambda_{13} \\ \lambda_{12} & 0 & 0 & 0 \\ 0 & \lambda_{13} & 0 & 0 \end{pmatrix},$$

and $U$ is the matrix that diagonalizes $\Lambda$. The eigenvalues of the matrix $\Lambda$ are solutions of the characteristic polynomial

$$\lambda^4 - (\lambda_1^2 + \lambda_{12}^2 + \lambda_{13}^2)\lambda^2 + \lambda_{12}^2\lambda_{13}^2 = 0, \tag{3.17}$$

that is,

$$\lambda_{\pm,\pm} = \pm\sqrt{(\lambda_1^2 + \lambda_{12}^2 + \lambda_{13}^2) \pm \sqrt{(\lambda_1^2 + \lambda_{12}^2 + \lambda_{13}^2)^2 - 4\lambda_{12}^2\lambda_{13}^2}}.$$

As it is trivially

$$\frac{\lambda_{+,\pm}}{\lambda_{-,\pm}} = -1,$$

we have that the commensurability of all the eigenvalues is guaranteed if and only if

$$\frac{\lambda_{+,+}}{\lambda_{-,-}} = \frac{p}{q},$$

with $p$ and $q$ being nonvanishing positive integers, that is, if and only if the condition

$$\frac{\lambda_1^2 + \lambda_{12}^2 + \lambda_{13}^2}{\lambda_{12}\lambda_{13}} = \frac{p}{q} + \frac{q}{p} \tag{3.18}$$

holds true.

Therefore, by computing $\hat{n}_{12}(t)$, $\hat{n}_{13}(t)$, $\hat{n}_2(t)$, and $\hat{n}_3(t)$, as well as their mean values on a state $\omega_{n_{12},n_{13},n_2,n_3}$ which generalizes the vector states introduced in Section 3.2, we can see that the solutions are, in general, quasiperiodic with two periods, and become periodic if Equation 3.18 is satisfied.

In Figure 3.5, we plot the solutions in the periodic case for $L_1 = L_2 = 3$, $n_{12}(0) = 0$, $n_{13}(0) = 3$, $n_2(0) = n_3(0) = 2$: Bob is strongly attracted by Carla, and he just does not care about Alice, whereas both Carla and Alice experience the same LoA for Bob. The parameters involved are the

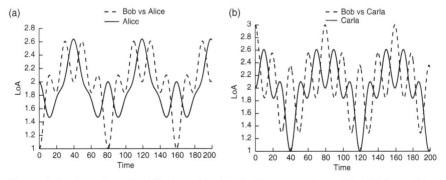

**Figure 3.5** $L_1 = L_2 = 3$: Alice's and Bob's LoAs versus time with initial condition $(0, 2)$ (a) and Carla's and Bob's LoA's versus time with initial condition $(2, 3)$ (b). Periodic behaviors are observed.

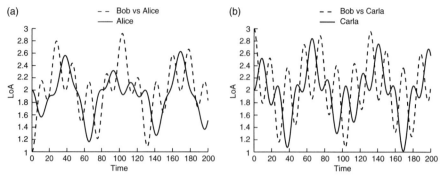

**Figure 3.6** $L_1 = L_2 = 3$: Alice's and Bob's LoA's versus time with initial condition $(0, 2)$ (a) and Carla's and Bob's LoA's versus time with initial condition $(2, 3)$ (b). Quasiperiodic behaviors are observed.

following: $\lambda_{12} = \frac{1}{10}$, $\lambda_{13} = \frac{1}{8}$, and $\lambda_1 = \frac{3}{40}$, which satisfies Equation 3.18 for $p = 2$ and $q = 1$.

The periodic behavior is clearly evident in both these plots. On the contrary, in Figure 3.6, we plot the solutions in the quasiperiodic case for the same $L_1$, $L_2$, $n_{12}(0)$, $n_{13}(0)$, $n_2(0)$, and $n_3(0)$ as earlier. The values of the parameters are now $\lambda_{12} = \frac{1}{10}$, $\lambda_{13} = \frac{1}{8}$, and $\lambda_1 \simeq 0.0889$, which satisfy Equation 3.18 for $p = \sqrt{5}$ and $q = 1$.

### 3.3.1 Another Generalization

A natural way to extend our previous Hamiltonian consists, in analogy with what we have done in Subsection 3.2.1, in introducing two different parameters $M_\alpha$, $\alpha = 12, 13$, which are able to describe different (relative) reactions in the two interactions Alice–Bob and Carla–Bob. We also assume that Bob is not very interested in choosing Carla rather than Alice, as long as one of the two is attracted by him. For these reasons, using **Rule 1** given earlier, the Hamiltonian of the triangle now looks like

$$H = \lambda_{12} \left( (a_{12}^\dagger)^{M_{12}} a_2 + a_{12}^{M_{12}} a_2^\dagger \right) + \lambda_{13} \left( (a_{13}^\dagger)^{M_{13}} a_3 + a_{13}^{M_{12}} a_3^\dagger \right)$$
$$+ \lambda_1 \left( a_{12}^\dagger a_{13} + a_{12} a_{13}^\dagger \right).$$

In addition, in this case, an integral of motion does exist

$$\tilde{J} := \hat{n}_{12} + \hat{n}_{13} + M_{12}\hat{n}_2 + M_{13}\hat{n}_3,$$

which reduces to $J$ if $M_{12} = M_{13} = 1$.

The equations of motion also extend to those given in Equation 3.15:

$$\begin{cases} i\,\dot{a}_{12}(t) = \lambda_{12}\,M_{12}\,(a_{12}^{\dagger}(t))^{M_{12}-1}a_2(t) + \lambda_1\,a_{13}(t), \\ i\,\dot{a}_{13}(t) = \lambda_{13}\,M_{13}\,(a_{13}^{\dagger}(t))^{M_{13}-1}\,a_3(t) + \lambda_1\,a_{12}(t), \\ i\,\dot{a}_2(t) = \lambda_{12}\,a_{12}(t), \\ i\,\dot{a}_3(t) = \lambda_{13}\,a_{13}(t). \end{cases} \tag{3.19}$$

The system of differential equation in (3.19) is nonlinear and cannot be solved analytically except when $M_{12} = M_{13} = 1$, as shown before. In fact, in this case, the system becomes linear and coincides with the one in Equation 3.15. Therefore, we adopt the same approach that we considered before, that is, we replace the *original* Hilbert space $\mathcal{H}$ with its effective, finite-dimensional version, $\mathcal{H}_{\text{eff}}$, whose dimensionality is fixed by the value of the integral of motion $J(t)$ at $t = 0$.

In particular, we take $L_1 = L_2 = 2$, and we consider the o.n. basis $\mathcal{F} = \{\varphi_{ijkl}\}$, where the indices $i, j, k, l$ run over the values $0,1,2$ in such a way that the sequence of four-digit numbers $ijkl$ (in base-3 numeral system) is sorted in ascending order. Therefore, $\dim(\mathcal{H}_{\text{eff}}) = (L_1 + 1)^2(L_2 + 1)^2 = 81$. In such a situation, the unknowns $a_{12}(t)$, $a_{13}(t)$, $a_2(t)$, and $a_3(t)$ in Equation 3.19 are replaced by square matrices of dimension 81, $A_{12}(t)$, $A_{13}(t)$, $A_2(t)$, and $A_3(t)$, which, at $t = 0$, look like:

$$A_{12}(0) = \begin{pmatrix} \mathbf{0}_{27} & \mathbf{0}_{27} & \mathbf{0}_{27} \\ \mathbb{1}_{27} & \mathbf{0}_{27} & \mathbf{0}_{27} \\ \mathbf{0}_{27} & \sqrt{2}\,\mathbb{1}_{27} & \mathbf{0}_{27} \end{pmatrix}, \quad A_{13}(0) = \begin{pmatrix} \underline{x} & \mathbf{0}_{27} & \mathbf{0}_{27} \\ \mathbf{0}_{27} & \underline{x} & \mathbf{0}_{27} \\ \mathbf{0}_{27} & \mathbf{0}_{27} & \underline{x} \end{pmatrix},$$

$$A_2(0) = \begin{pmatrix} \underline{y} & \mathbf{0}_{27} & \mathbf{0}_{27} \\ \mathbf{0}_{27} & \underline{y} & \mathbf{0}_{27} \\ \mathbf{0}_{27} & \mathbf{0}_{27} & \underline{y} \end{pmatrix}, \quad A_3(0) = \begin{pmatrix} \underline{z} & \mathbf{0}_{27} & \mathbf{0}_{27} \\ \mathbf{0}_{27} & \underline{z} & \mathbf{0}_{27} \\ \mathbf{0}_{27} & \mathbf{0}_{27} & \underline{z} \end{pmatrix},$$

where

$$\underline{x} = \begin{pmatrix} \mathbf{0}_9 & \mathbf{0}_9 & \mathbf{0}_9 \\ \mathbb{1}_9 & \mathbf{0}_9 & \mathbf{0}_9 \\ \mathbf{0}_9 & \sqrt{2}\,\mathbb{1}_9 & \mathbf{0}_9 \end{pmatrix}, \quad \underline{y} = \begin{pmatrix} \underline{y}_1 & \mathbf{0}_9 & \mathbf{0}_9 \\ \mathbf{0}_9 & \underline{y}_1 & \mathbf{0}_9 \\ \mathbf{0}_9 & \mathbf{0}_9 & \underline{y}_1 \end{pmatrix},$$

$$\underline{y}_1 = \begin{pmatrix} \mathbf{0}_3 & \mathbf{0}_3 & \mathbf{0}_3 \\ \mathbb{1}_3 & \mathbf{0}_3 & \mathbf{0}_3 \\ \mathbf{0}_3 & \sqrt{2}\,\mathbb{1}_3 & \mathbf{0}_3 \end{pmatrix}, \quad \underline{z} = \begin{pmatrix} \underline{z}_1 & \mathbf{0}_9 & \mathbf{0}_9 \\ \mathbf{0}_9 & \underline{z}_1 & \mathbf{0}_9 \\ \mathbf{0}_9 & \mathbf{0} & \underline{z}_1 \end{pmatrix},$$

$$\underline{z}_1 = \begin{pmatrix} \underline{z}_2 & \mathbf{0}_3 & \mathbf{0}_3 \\ \mathbf{0}_3 & \underline{z}_2 & \mathbf{0}_3 \\ \mathbf{0}_3 & \mathbf{0}_3 & \underline{z}_2 \end{pmatrix}, \quad \underline{z}_2 = \begin{pmatrix} 0 & 0 & 0 \\ 1 & 0 & 0 \\ 0 & \sqrt{2} & 0 \end{pmatrix}.$$

Moreover, in order to consider a situation not very far from a linear one, we choose $M_{12} = 1$, $M_{13} = 2$, $\lambda_1 = 0.3$, $\lambda_{12} = 0.2$, $\lambda_{13} = 0.007$.

The numerical integration has been performed by using once again the ode113 routine of MATLAB®, and the existence of the integral of motion has been used in order to check the accuracy of the solutions.

The choice $\lambda_{13} = 0.007$ has an immediate consequence, which is clearly displayed by the plots in Figures 3.7–3.10. The time evolution of $\hat{n}_3(t)$ appears essentially trivial in the time interval where the system is numerically integrated. Carla keeps her initial status or, stated with different words, *her feelings are stable*. This does not imply, however, that Bob's LoA for Carla stays constant as well, as one can deduce from the same figures. This means that whereas the value of $\lambda_{13}$ determines Carla's feelings, Bob's behavior is related to $\lambda_{12} + \lambda_{13}$, which is, in a sense, expected. Another interesting feature, which is again due to the nonlinear aspect of the dynamics (and it was not present in the linear model), is related to the fact that, in certain time intervals, Bob's LoA for both Alice and Carla can increase. This is, in a certain sense, surprising: the Hamiltonian does not present any explicit ingredient responsible for this feature, which, in

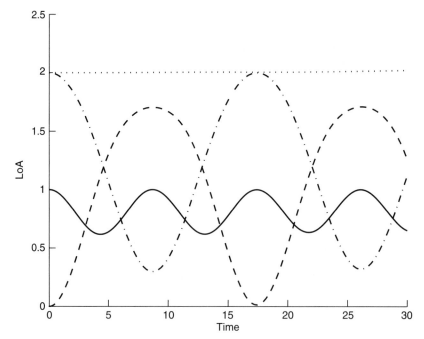

**Figure 3.7**    $L_\alpha = 2$, $M_{12} = 1$, $M_{13} = 2$ ($\alpha = 12, 13, 2, 3$): LoA versus time of Bob versus Alice (continuous line), Bob versus Carla (dashed line), Alice (dashed-dotted line), and Carla (dotted line) with initial condition $(1, 0, 2, 2)$. The plots show periodic behaviors.

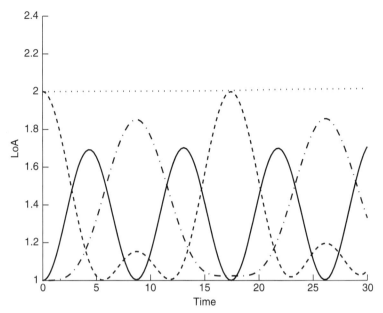

**Figure 3.8**   $L_\alpha = 2$, $M_{12} = 1$, $M_{13} = 2$ ($\alpha = 12, 13, 2, 3$): LoA versus time of Bob versus Alice (continuous line), Bob versus Carla (dashed line), Alice (dashed-dotted line), and Carla (dotted line) with initial condition $(1, 2, 1, 2)$. The plots show periodic behaviors.

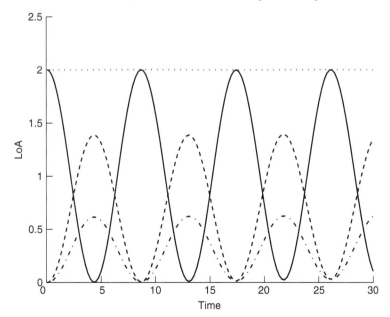

**Figure 3.9**   $L_\alpha = 2$, $M_{12} = 1$, $M_{13} = 2$ ($\alpha = 12, 13, 2, 3$): LoA versus time of Bob versus Alice (continuous line), Bob versus Carla (dashed line), Alice (dashed–dotted line), and Carla (dotted line) with initial condition $(2, 0, 0, 2)$. The plots show periodic behaviors.

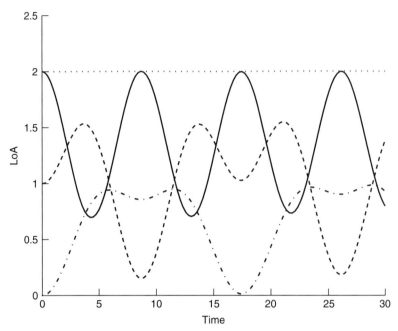

**Figure 3.10**   $L_\alpha = 2$, $M_{12} = 1$, $M_{13} = 2$ ($\alpha = 12, 13, 2, 3$): LoA versus time of: Bob versus Alice (continuous line), Bob versus Carla (dashed line), Alice (dashed-dotted line), and Carla (dotted line) with initial condition $(2, 1, 0, 2)$. The plots show periodic behaviors.

our opinion, appears just because of the nonlinearity of the interaction. Finally, we also notice that Figures 3.7–3.10 display a periodic behavior, even if no analytical proof of the existence of this periodicity is still available.

A crucial remark, which was discussed first in Section 1.3 and which we are checking here in our explicit models, is that despite the quantum framework adopted in this chapter, all the observables relevant to the description of the models introduced so far commute among them and, for this reason, can be simultaneously measured: no Heisenberg uncertainty principle has to be invoked for these operators; from this point of view, there is no difference between a *standard classical system* and the system constructed in this section.

Of course, this would not be true, in general, if we were interested in computing two quantities both related to a single actor, for instance, to Bob. Suppose, in fact, that we want to compute, for some reason, the mean values of the following self-adjoint operators:

$$x_{13} := \frac{1}{\sqrt{2}} \left( a_{13} + a_{13}^\dagger \right), \qquad p_{13} := \frac{1}{\sqrt{2}\,i} \left( a_{13} - a_{13}^\dagger \right),$$

which in ordinary quantum mechanics are, respectively, the position and the momentum operators. Then, as $[x_{13}, p_{13}] = i\,\mathbb{1}$, these two operators must obey the Heisenberg uncertainty principle as in Section 2.4, and they cannot both be measured with arbitrary precision.

We end this section with two final comments: the first is related to the oscillatory behavior (periodic or quasiperiodic) recovered for the different LoAs of our actors. Indeed, this is expected, because of the simple Hamiltonian considered here. In the next section, we show how some time decay in the LoAs can be recovered by means of possible interactions of the lovers with the *environment*. Our second comment concerns the fact that no *free* Hamiltonian, $H_0$, has been considered to deduce the dynamics here. Our future results will show that this could be really a limitation of the model as the parameters appearing in $H_0$ usually play an interesting role in the dynamics of each interacting system. Adding $H_0$ to the interaction Hamiltonians considered so far is an exercise that we leave to the curious reader, even because we will consider such a contribution in the next section, in a slightly more general situation.

## 3.4 DAMPED LOVE AFFAIRS

We now describe how to extend the general strategy discussed in Section 3.2 to produce a model where more effects, and not only mutual interaction between Bob and Alice, can be taken into account. Suppose that Alice and Bob interact with their own *reservoirs* (Bagarello, 2011). We call $\mathcal{S} = \mathcal{S}_a \cup \mathcal{S}_b \cup \mathcal{R}_A \cup \mathcal{R}_B$ the full system, made of Alice ($\mathcal{S}_a$), Bob ($\mathcal{S}_b$), Alice's reservoir ($\mathcal{R}_A$), and Bob's reservoir ($\mathcal{R}_B$). We could think of these reservoirs as the *real world* surrounding Alice and Bob: their friends, enemies, families,. ... For the sake of simplicity, we forget here about the *disturbing* effects produced by Carla. Extending the Hamiltonian introduced in Equation 2.67 in a more abstract situation to the present settings, we define

$$\begin{cases} H = H_A + H_B + \lambda H_I, \\ H_A = \omega_a a^\dagger a + \int_{\mathbb{R}} \Omega_A(k) A^\dagger(k) A(k)\, \mathrm{d}k \\ \qquad + \gamma_A \int_{\mathbb{R}} \left( a^\dagger A(k) + a A^\dagger(k) \right) \mathrm{d}k, \\ H_B = \omega_b b^\dagger b + \int_{\mathbb{R}} \Omega_B(k) B^\dagger(k) B(k)\, \mathrm{d}k \\ \qquad + \gamma_B \int_{\mathbb{R}} \left( b^\dagger B(k) + b B^\dagger(k) \right) \mathrm{d}k, \\ H_I = a^\dagger b + a b^\dagger \end{cases} \qquad (3.20)$$

The constants $\omega_a$, $\omega_b$, $\gamma_a$, $\gamma_b$, and $\lambda$ are all real to ensure that $H = H^\dagger$. Recall that a regularization should also be considered in Equation 3.20 to make the Hamiltonian rigorously defined. We will skip these mathematical details here as we are more interested in the physical meaning of $H$ and as, using, for instance, the SLA sketched in Section 2.5.4, we could make our treatment rigorous anyhow. The following bosonic commutation rules are assumed:

$$[a, a^\dagger] = [b, b^\dagger] = \mathbb{1}, \quad [A(k), A^\dagger(q)] = [B(k), B^\dagger(q)] = \mathbb{1}\delta(k - q),$$
$$(3.21)$$

whereas all the other commutators are zero. $H_A$ and $H_B$ are copies of the Hamiltonian in Equation 2.67. They describe respectively the interaction of Alice and Bob with their own reservoirs, which are made of several (actually infinite) *ingredients*, which, as we have already said, can be interpreted as other men, other women, and also as the effect of tiredness, tedium, and so on. $H_I$ describes the interaction between Alice and Bob. We see that this is nothing but a linear version of the interaction Hamiltonian originally given in Equation 3.5. We also stress that as already mentioned at the end of the previous section, Equation 3.20 for $H$ also contains a *free* Hamiltonian,

$$H_0 = \omega_a a^\dagger a + \omega_b b^\dagger b + \int_{\mathbb{R}} \Omega_A(k) A^\dagger(k) A(k)\, dk$$

$$+ \int_{\mathbb{R}} \Omega_B(k) B^\dagger(k) B(k)\, dk. \tag{3.22}$$

The reasons this particular expression for $H_0$ was chosen are discussed later on in this section and, in more detail, in Chapter 6. Here, we just want to mention that this is the *canonical* aspect of a free Hamiltonian for (almost) all second-quantized systems.

The Heisenberg equations of motion for the annihilation operators are the following:

$$\begin{cases} \dot{a}(t) = i[H, a(t)] = -i\omega_a a(t) - i\gamma_A \int_{\mathbb{R}} A(k, t)\, dk - i\lambda b(t), \\ \dot{b}(t) = i[H, b(t)] = -i\omega_b b(t) - i\gamma_B \int_{\mathbb{R}} B(k, t)\, dk - i\lambda a(t), \\ \dot{A}(k, t) = i[H, A(k, t)] = -i\Omega_A(k) A(k, t) - i\gamma_A a(t), \\ \dot{B}(k, t) = i[H, B(k, t)] = -i\Omega_B(k) B(k, t) - i\gamma_B b(t). \end{cases} \tag{3.23}$$

Using the same strategy discussed in Section 2.7, and therefore fixing $\Omega_A(k) = \Omega_A k$ and $\Omega_B(k) = \Omega_B k$, $\Omega_A, \Omega_B > 0$, we further get

$$\begin{cases} \dot{a}(t) = -\nu_A a(t) - i\lambda b(t) - i\gamma_A f_A(t), \\ \dot{b}(t) = -\nu_B b(t) - i\lambda a(t) - i\gamma_B f_B(t), \end{cases} \tag{3.24}$$

where

$$\nu_A = \frac{\pi \gamma_A^2}{\Omega_A} + i\omega_A, \qquad \nu_B = \frac{\pi \gamma_B^2}{\Omega_B} + i\omega_B,$$

and

$$f_A(t) = \int_{\mathbb{R}} e^{-i\Omega_A k t} A(k)\, dk, \qquad f_B(t) = \int_{\mathbb{R}} e^{-i\Omega_B k t} B(k)\, dk.$$

From Equation 3.24, we deduce the following second-order differential equation for $a(t)$:

$$\ddot{a}(t) + \dot{a}(t)(\nu_A + \nu_B) + a(t)(\nu_A \nu_B + \lambda^2) = \Phi(t), \tag{3.25}$$

where $\Phi(t) := -i\gamma_A \dot{f}_A(t) - i\nu_B \gamma_A f_A(t) - \lambda \gamma_B f_B(t)$. A similar second-order differential equation could also be deduced for $b(t)$. However, for reasons that will be clear in a moment, this other equation is not needed. To find the solution of Equation 3.25, satisfying $a(0) = a$, we need to compute the Green's function for the differential operators $L := \frac{d^2}{dt^2} + (\nu_A + \nu_B)\frac{d}{dt} + (\nu_A \nu_B + \lambda^2)$, that is, the function $G(t)$ satisfying $L[G](t) = \delta(t)$, see Section 2.8. To keep the computations reasonably simple from now on, we fix $\nu_A = \nu_B =: \nu$, which implies that $\frac{\gamma_A^2}{\Omega_A} = \frac{\gamma_B^2}{\Omega_B}$, and $\omega_A = \omega_B =: \omega$. It is a standard exercise in Fourier transform to deduce that

$$G(t) = \begin{cases} \dfrac{1}{\lambda} \sin(\lambda t)\, e^{-\nu t}, & t > 0 \\ 0, & \text{otherwise.} \end{cases}$$

The general solution of the equation $L[a_0(t)] = 0$ is $a_0(t) = x_+ e^{\epsilon_+ t} + x_- e^{\epsilon_- t}$, with $\epsilon_\pm = (-\nu \pm i\lambda)$. Noting that $b(t)$ can be deduced from $a(t)$ by means of the first equation in 3.24, which can be rewritten as $b(t) = \frac{1}{\lambda}(i\dot{a}(t) + i\nu_A a(t) - \gamma_A f_A(t))$, we find that

$$\begin{cases} a(t) = ae^{-\nu t}\cos(\lambda t) - ibe^{-\nu t}\sin(\lambda t) + R_a(t), \\ b(t) = be^{-\nu t}\cos(\lambda t) - iae^{-\nu t}\sin(\lambda t) + R_b(t), \end{cases} \tag{3.26}$$

where we have defined the following functions:

$$
\begin{cases}
R_a(t) = \rho(t) + \dfrac{e^{-\nu t}}{\lambda}\left\{ i\Gamma(0)\sin(\lambda t) - \lambda\rho(0)\cos(\lambda t) \right\}, \\[2mm]
R_b(t) = \dfrac{1}{\lambda}\left\{ \Gamma(t) - \Gamma(0)e^{-\nu t}\cos(\lambda t) \right\} + i\rho(0)e^{-\nu t}\sin(\lambda t), \\[2mm]
\rho(t) = \int_{\mathbb{R}} \left(\rho_A(k,t)A(k) + \rho_B(k,t)B(k)\right)\,dk, \\[2mm]
\Gamma(t) = \int_{\mathbb{R}} \left(\Gamma_A(k,t)A(k) + \Gamma_B(k,t)B(k)\right)\,dk.
\end{cases}
\tag{3.27}
$$

Here, we have defined:

$$
\begin{cases}
\rho_A(k,t) = -\gamma_A \dfrac{\Omega_A k - i\nu}{\lambda^2 + (i\Omega_A k - \nu)^2}\,e^{-i\Omega_A kt}, \\[3mm]
\rho_B(k,t) = \dfrac{-\lambda\gamma_B}{\lambda^2 + (i\Omega_B k - \nu)^2}\,e^{-i\Omega_B kt}, \\[3mm]
\Gamma_A(k,t) = i\dot{\rho}_A(k,t) + i\nu\rho_A(k,t) - \gamma_A e^{-i\Omega_A kt}, \\[2mm]
\Gamma_B(k,t) = i\dot{\rho}_B(k,t) + i\nu\rho_B(k,t).
\end{cases}
$$

It is now easy to find the mean value of the number operators $\hat{n}_a(t) = a^\dagger(t)a(t)$ and $\hat{n}_b(t) = b^\dagger(t)b(t)$ on a state over $\mathcal{S}$, which is of the form $\omega_S(.) := \langle \varphi_{n_a,n_b}, \cdot\, \varphi_{n_a,n_b}\rangle \omega_{\mathcal{R}}(.)$. Here $\omega_{\mathcal{R}}$ is a suitable state over the reservoirs $\mathcal{R} = \mathcal{R}_A \cup \mathcal{R}_B$, whereas, as usual, $\varphi_{n_a,n_b}$ is the eigenstate of $\hat{n}_a$ and $\hat{n}_b$ with eigenvalues $n_a$ and $n_b$. Then, calling $n_a(t) = \omega_S(\hat{n}_a(t))$ and $n_b(t) = \omega_S(\hat{n}_b(t))$, and assuming that, see Section 2.7,

$$
\begin{aligned}
\omega_{\mathcal{R}}(A^\dagger(k)A(q)) &= N_A(k)\delta(k-q), \\
\omega_{\mathcal{R}}(B^\dagger(k)B(q)) &= N_B(k)\delta(k-q),
\end{aligned}
\tag{3.28}
$$

we conclude that

$$
n_a(t) = e^{-2\pi\gamma_A^2 t/\Omega_A}\left(n_a\cos^2(\lambda t) + n_b\sin^2(\lambda t)\right)
$$
$$
+ \int_{\mathbb{R}}\left(N_A(k)|\mu_{a,A}(k,t)|^2 + N_B(k)|\mu_{a,B}(k,t)|^2\right)\,dk
\tag{3.29}
$$

and

$$
n_b(t) = n_b\,e^{-2\pi\gamma_A^2 t/\Omega_A}\cos^2(\lambda t) + n_a\,e^{-2\pi\gamma_A^2 t/\Omega_A}\sin^2(\lambda t)
$$
$$
+ \int_{\mathbb{R}}\left(N_A(k)|\mu_{b,A}(k,t)|^2 + N_B(k)|\mu_{b,B}(k,t)|^2\right)\,dk,
\tag{3.30}
$$

where we have introduced the following functions:

$$\mu_{a,A}(k,t) = \rho_A(k,t) + \frac{i}{\lambda}e^{-\nu t}\sin(\lambda t)\Gamma_A(k,0) - e^{-\nu t}\cos(\lambda t)\rho_A(k,0),$$

$$\mu_{a,B}(k,t) = \rho_B(k,t) + \frac{i}{\lambda}e^{-\nu t}\sin(\lambda t)\Gamma_B(k,0) - e^{-\nu t}\cos(\lambda t)\rho_B(k,0),$$

$$\mu_{b,A}(k,t) = \frac{1}{\lambda}\Gamma_A(k,t) - \frac{1}{\lambda}e^{-\nu t}\cos(\lambda t)\Gamma_A(k,0) - e^{-\nu t}\sin(\lambda t)\rho_A(k,0),$$

$$\mu_{b,B}(k,t) = \frac{1}{\lambda}\Gamma_B(k,t) - \frac{1}{\lambda}e^{-\nu t}\cos(\lambda t)\Gamma_B(k,0) - e^{-\nu t}\sin(\lambda t)\rho_B(k,0).$$

Equations 3.29 and 3.30 look interesting as they clearly display the different contributions arising from the system $\mathcal{S}_a \cup \mathcal{S}_b$ and from the reservoir $\mathcal{R}$. Of course, different choices of the functions of the reservoir $N_A(k)$ and $N_B(k)$ clearly produce different expressions for the LoAs of Alice and Bob. In particular, if, for instance, $N_A(k) = N_B(k) = 0$ almost everywhere in $k$, we simply get

$$\begin{cases} n_a(t) = e^{-2\pi\gamma_A^2 t/\Omega_A}\left(n_a\cos^2(\lambda t) + n_b\sin^2(\lambda t)\right), \\ n_b(t) = e^{-2\pi\gamma_A^2 t/\Omega_A}\left(n_b\cos^2(\lambda t) + n_a\sin^2(\lambda t)\right), \end{cases} \tag{3.31}$$

which represent damped oscillations for both Alice and Bob: independently of the initial conditions, the effect of the reservoirs is to switch off the love between Alice and Bob, at least for this trivial choice of $N_A(k)$ and $N_B(k)$. In this case, the reservoirs are both empty at $t = 0$: they behave as two sinks absorbing all the love that Alice and Bob originally experienced for each other. This conclusion is justified later. Going back to Equation 3.31, if $n_a = n_b = n$, that is, if Alice and Bob experience the same LoA at $t = 0$, we get $n_a(t) = n\,e^{-2\pi\gamma_A^2 t/\Omega_A}$ and $n_b(t) = n\,e^{-2\pi\gamma_A^2 t/\Omega_A}$. The speed of decay of their LoA is related to $\gamma_A^2/\Omega_A$ which, we recall, coincides in this case, because of our assumptions, with $\gamma_B^2/\Omega_B$. In particular, the stronger the interaction between, say, Alice and her reservoir, the faster her love for Bob goes to zero. Of course, a different speed is expected, in general, for Bob's LoA for Alice. This, we believe, could be recovered if $\nu_A \neq \nu_B$. In this case, we expect different asymptotic behaviors for Alice and Bob. However, this generalization is not considered here.

We briefly discuss the case in which $N_A(k)$ and $N_B(k)$ are different from 0 in the next section. Now we want to comment on a *new* feature of our model. In contrast to what we have done in the first part of this chapter, we have considered here the free part of the Hamiltonian $H$, $H_0$. So one may wonder why this is done here and why it is not enough to consider again a simplified Hamiltonian obtained from Equation 3.20 by taking $\omega_a = \omega_b = \Omega_A(k) = \Omega_B(k) = 0$. In this case, the differential equations for the annihilation operators of the system and the reservoir are easily deduced but the solution is purely oscillatory: if we do not consider the free Hamiltonian of the reservoirs, the solution is a finite combination of periodic functions: no damping seems to be possible. This is not the only reason the free Hamiltonian plays a relevant role. We will see many times in the subsequent chapters that changing the parameters of $H_0$, some features of the dynamics of the system significantly change as well. This will allow us to give a *physical* meaning to the parameters of $H_0$. Furthermore, concerning the way in which $H_0$ is chosen, see Equation 3.22 for instance, in these notes we adopt everywhere a sort of *part 2* of the **Rule 1**, introduced in Section 3.3:

**Rule 2:** *The Hamiltonian $H_0$ of $S$ is such that, in absence of interactions between the elements of $S$, their number operators stay constant in time.*

For instance, going back to the love triangle considered in this chapter, the natural choice would be $H_0 = \omega_1 a_1^\dagger a_1 + \omega_2 a_2^\dagger a_2 + \omega_3 a_3^\dagger a_3$ for some real $\omega_1, \omega_2$, and $\omega_3$. Indeed, with this choice, it is clear that when Alice, Bob, and Carla do not interact (all the $\lambda$s are zero), $H \equiv H_0$ and $\hat{n}_j(t) = \hat{n}_j(0)$ for all $j$. This is a simple consequence of the following commutators: $[H, \hat{n}_1] = [H, \hat{n}_2] = [H, \hat{n}_3] = 0$.

### 3.4.1 Some Plots

Let us now consider what happens if $N_A(k)$ and $N_B(k)$ are different from zero. First, we consider the following case: $N_A(k) = N_B(k) = 5\,e^{-k^2}$, where 5 is chosen because of our initial conditions on $n_a$ and $n_b$, which we assume, to fix the ideas, can only take values 1 and 5: $n_a = 1$ and $n_b = 5$, see Figure 3.11, corresponds to a low Alice's LoA and a high Bob's LoA. In other words, we do not want to have huge differences between $S$ and $R$. The other parameters take the following values: $\omega = 1$, $\Omega_A = \Omega_B = 1$, $\gamma_A = \gamma_B = 0.1$, and $\lambda = 0.3$. So the situation looks rather symmetrical, and the only asymmetry is given by the initial conditions mentioned earlier.

These figures show nonpurely oscillatory behaviors of $n_a(t)$ and $n_b(t)$, which seem to converge to a limiting asymptotic value for $t$ large enough: this is like a sort of reasonable *stationary state* in which both Bob and Alice will continue their relationship with not many changes in their LoAs. The reservoir acts as a pump, giving stability to their love relation.

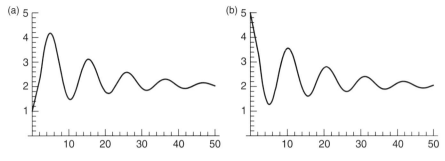

**Figure 3.11** $n_a(t)$ (a) and $n_b(t)$ (b).

However, this is not always like this. First, if we just change the initial conditions requiring that $n_a = n_b = 5$; it is easy to see that $n_a(t) = n_b(t)$ and that they both decay monotonically to a value close to 2.5, which apparently seems to be close to the asymptotic value deduced from Figure 3.11.

On the contrary, if we now break down the original symmetry between Alice and Bob not only considering different initial conditions but also changing the values of the parameters, then the asymptotic behavior of $n_a(t) = n_b(t)$ is not so clear, see Figure 3.12. These plots are obtained taking $\omega$ and $\lambda$ as earlier, $\Omega_A = 1$ $\Omega_B = 4$, $\gamma_A = 0.1$ and $\gamma_B = 0.2$. This choice satisfies the requirement $\gamma_A^2/\Omega_A = \gamma_B^2/\Omega_B$.

We see from the figure that $n_a(t)$ and $n_b(t)$ do not necessarily tend to a certain value for increasing $t$, at least in the time interval considered in these plots. In these conditions it is quite hard to find an equilibrium! At least, it takes longer. Of course this means, because of Equations 3.29 and 3.30, that the presence of the reservoir might cause these smaller *oscillations* of the original functions. Similar plots are obtained if we increase further the asymmetry between Alice and Bob by *attaching* them to two different reservoirs. Indeed, if we take now $N_A(k) = 5\,e^{-k^2}$ and

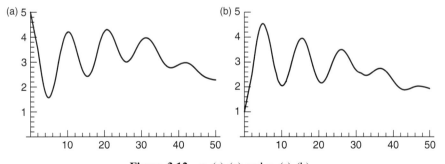

**Figure 3.12** $n_a(t)$ (a) and $n_b(t)$ (b).

$N_B(k) = 5\,e^{-3k^2}$, leaving all the other constants unchanged, we get the functions plotted in Figure 3.13, for $n_a = 1$ and $n_b = 5$; in Figure 3.14, for $n_a = 5$ and $n_b = 1$; and in Figure 3.15, for $n_a = 5$ and $n_b = 5$.

In particular, Figure 3.15 shows that both $n_a(t)$ and $n_b(t)$ decrease, not monotonically, and that some limiting value, if any, is reached very far

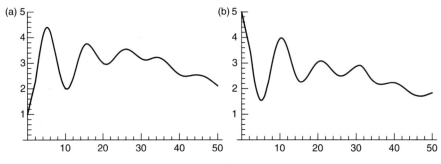

**Figure 3.13**   $n_a(t)$ (a) and $n_b(t)$ (b).

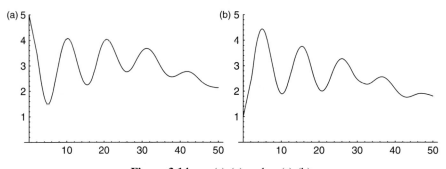

**Figure 3.14**   $n_a(t)$ (a) and $n_b(t)$ (b).

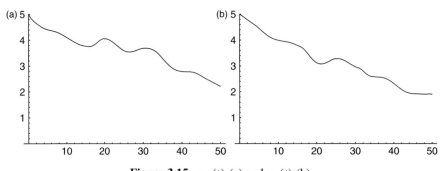

**Figure 3.15**   $n_a(t)$ (a) and $n_b(t)$ (b).

away (compared with what happens in Figure 3.11) in time. Moreover, the same figure also clearly displays that $n_a(t) + n_b(t)$ is not a constant of motion, as expected because the contributions of the reservoirs are neglected: however, even for this model, an integral of motion still exists, and can be written as:

$$\tilde{I}(t) = \hat{n}_a(t) + \hat{n}_b(t) + \int_{\mathbb{R}} dk\, A^{\dagger}(k,t) A(k,t) + \int_{\mathbb{R}} dk\, B^{\dagger}(k,t) B(k,t).$$

Hence, as we have already suggested before, what is lost by the system is acquired by the reservoir, and vice versa: if Bob loses interest in Alice, this is because, quite reasonably, he acquires interest in somebody else. Considering again Figure 3.15, we also see that the presence of the reservoir is not enough to get a substantially different behavior of Alice and Bob if they start with the same LoA. This is probably due to the fact that the analytic expressions for $N_A(k)$ and $N_B(k)$ that have been chosen here are very similar: stronger differences in $n_a(t)$ and $n_b(t)$ possibly require major differences in the reservoirs. Of course, this is physically motivated: if Alice and Bob have different *inputs*, it is natural to expect that they behave differently, although there is no reason, *a priori*, for them to show different dynamics if they share similar interactions with the rest of the world.

It is clear that the possibility of having damping effects makes the model more interesting and realistic, also in view of the *physical* interpretation of the reservoir, which plays the role of the world surrounding Alice and Bob, and which makes these two guys a part of the reality, considered here as an open system, rather than as the closed system we have described in Sections 3.2 and 3.3.

It is also clear from our figures that the value of the parameters of $H_A + H_B$ may really change the details of the dynamics of the system, leaving unchanged the general structure of the solution: whereas the choice of parameters adopted in Figure 3.11 produces a sort of equilibrium quite fast, the other choices do not: we have to wait much longer for such an equilibrium between Alice and Bob to be reached!

Hence, the free Hamiltonian is the one that determines the time interval $[0, t_c]$ in which relevant variations in the love affair may take place. For $t > t_c$ the changes in the LoAs of both lovers are extremely small. Not surprisingly, the value of $t_c$ depends in an essential way from the strength of the interaction of Alice and Bob with their reservoirs, that is, from $\gamma_a$ and $\gamma_b$: the stronger the interaction of, say, Bob, with the world around him (with other women, for instance), the faster he loses interest in Alice.

## 3.5   COMPARISON WITH OTHER STRATEGIES

This chapter described one of the few existing applications of mathematics to the description of love affairs. The first ancestor of these applications is probably the paper by Strogatz (1988), a very short note about Romeo and Juliet where a mechanism very similar to the one assumed here is considered: the more Juliet loves Romeo, the more he begins to dislike her. The model proposed is linear, giving rise to harmonic oscillations in the behavior of Romeo and Juliet. The one in Strogatz (1988) is just a very first step for further, and richer, extensions such as those described in Rinaldi (1998b) and Sproot (2004). In particular, the author in Rinaldi (1998b) tries to explain why two persons who are, at $t = 0$, completely indifferent to each other can develop a love affair. For that, the ingredients considered in the deduction of the dynamics are the *forgetting process*, the *pleasure of being loved*, and the *reaction to the appeal of the partner*. Rinaldi's interest is mainly concentrated on the conditions in which, starting from a complete indifference, a stable relationship is reached. In a certain sense, we have somehow answered this problem: Figure 3.11 suggests that, in presence of suitable reservoirs, and for particular initial values of the LoAs, such a stable relationship can be reached.[4] However, it seems more likely to find some decay in the feelings which, when the love story begins, are usually extremely strong. The model considered by Rinaldi is linear to simplify the analysis. Rinaldi also considers a generalization to the community level, when many people interact.

In Sproot (2004), love triangles and nonlinearities are considered, making the situation described there similar to what we have described here using some of the Hamiltonians introduced in this chapter. Sprott's paper begins with a more philosophical question: what is love and how can it be measured? This is far from being an easy question, and we give no answer at all. We end stressing once again that, although in Strogatz (1988), Rinaldi (1998b), and Sproot (2004), as well as in the interesting book by Gottman et al. (2002) love is a function of time whose evolution is driven by some differential equation, in this chapter, love is a time-dependent operator, whose eigenvalues measure its value, and the dynamics is driven by a self-adjoint Hamiltonian whose expression can be deduced from some general requirements, that is, by adopting **Rules 1** and **2** introduced in this chapter.

---

[4]To deal properly with the original Rinaldi's problem, we should consider what happens when we take $n_a = n_b = 0$ as this corresponds to a complete indifference of the lovers at $t = 0$.

# CHAPTER 4

# MIGRATION AND INTERACTION BETWEEN SPECIES

In this chapter, we consider an application of the CAR to a system consisting of two main (sets of) actors, representing two different biological species subjected to a mutual interaction. Most of this chapter is devoted to the analysis of a migration process, while in Section 4.5 we show how a simple change of the numerical values of the parameters of the model allows us to use the same Hamiltonian for a phenomenological description of competitions between species.

It is perhaps worth remarking that here we are considering only the very general features of a migration process or a predator–prey system. This was, essentially, what we also did in Chapter 3: the *details* of a love affair, as well as those of a, say, migration, are so many and involve so many variables that a comprehensive analysis is, at least for the moment, out of reach. However, it is interesting to notice that essentially our same framework has been widely used by Yukalov and his collaborators, in recent years, in connection with what they call *quantum decision theory* (Yukalov and Sornette, 2009a, b; Yukalov and Sornette). This topic is, clearly, very important in connection with almost all the systems we are considering in this book, and for this reason we plan to look, in the near future, for reasonable *intersections* between the two points of view.

*Quantum Dynamics for Classical Systems: With Applications of the Number Operator*,
First Edition. Fabio Bagarello.
© 2013 John Wiley & Sons, Inc. Published 2013 by John Wiley & Sons, Inc.

## 4.1 INTRODUCTION AND PRELIMINARIES

A large body of theoretical and experimental evidence that spatial patchy environments influence the dynamics of species interactions has been available since many years in the literature (Hanski, 1981, 1983, 2008; Ives and May, 1985; Comins and Hassel, 1987; Comins et al., 1992). Hence, a broad variety of spatially extended models has been developed in theoretical biology. The principle of competitive exclusion (Gause's law), stating that two like (identical) species cannot coexist in the same habitat, is violated in patchy environments in which two like species may coexist because of migration (Slatkin, 1974). A lot of evidence exists about the coexistence, in a fragmented environment, of two competing species (or populations) even if one is competitively superior to the other; see Hanski and Gilpin (1991), Nee and Mat (1992) and references therein. Besides the usual models based on continuous reaction–diffusion equations (Murray, 2003) and cellular automata, the coupled map lattice (CML) formalism has been widely used in the simulation of biological spatial interactions. In the usual CML approach, local prey–predator (or host–parasitoid) dynamics are coupled with their $n$–nearest neighbors through some appropriate exchange rule. Populations interact and disperse over the points of a two-dimensional lattice, which is used to simulate the patchy environment. In such a context, various aspects can be observed, such as the emergence of some persistent spatial patterns in the distributions of the competing species (e.g., phytoplankton distribution in the oceans), or the phenomena of synchronization between the phases of nearby regions (Oliveri and Paparella, 2008).

Different kinds of models of competing populations, also including spatial interactions, can be constructed using the same operator methods adopted in the previous chapter, and, in particular, assuming that the dynamics of the system $S$ can be described by an energy-like operator, the *Hamiltonian* of $S$, by means of the Heisenberg equation of motion 2.2. Using this strategy, we describe in Section 4.2 a (strictly local) situation in which two populations live together and are forced to interact, whereas in Section 4.3 we consider a second situation in which the two species occupy (in general) different cells of a two-dimensional lattice, interact, and move along the cells. These models can be useful for modeling different biological and/or sociological systems. In particular, we restrict ourselves here to consider the following two: a migration process in which a population moves from a given (poor) place to a richer region of the lattice, which is already occupied by a second group of people, and a system in which the two populations compete all over the lattice. In our opinion, it is particularly interesting to note that for the description of these two different systems, a single Hamiltonian $H$ is needed: it will become clear in the following discussion that these two situations can be recovered by some natural tuning of the parameters defining $H$.

Rather than the CCR, in this chapter, the relevant commutation rules are those described by the CAR. This choice is based mainly on two reasons. The first one is of a technical nature: the Hilbert spaces of our models are automatically finite dimensional, so that all the observables are bounded operators. In a certain sense, this is a simplified version of what we have done in Section 3.2.1, where we have introduced an approximation producing, out of $\mathcal{H}$, the finite-dimensional Hilbert space $\mathcal{H}_{\text{eff}}$: both procedures produce finite-dimensional Hilbert spaces, and bounded operators (more than this: finite-dimensional matrices!) as a consequence. However, while in Chapter 3 this was the effect of a suitable cutoff, here we do not make use of any approximation, making, in a certain sense, the derivation of the equations of motion more straightforward. The second reason is related to the biological and/or sociological interpretation of our model: for each population that we consider, we admit only two possible nontrivial situations. In the first one (the *ground state*), there is a very low density of the populations, while in the second (the *excited state*), the density is very high. Hence, if we try to increase the density of the excited state, or if we try to decrease the density of the ground state, we simply annihilate that population! We can interpret this fact by just saying that there exist upper and lower bounds for the densities of the populations that cannot be exceeded for obvious reasons: for instance, because the environment cannot give enough food to the populations. Of course, this rather sharp division in just two levels may appear unsatisfactory. However, it is not hard to extend our procedure to an arbitrary, and finite, number of levels, paying the price of some technical difficulties, mainly concerning the commutation rules that have to be changed. We do not consider this extension here as already in our hypotheses, an absolutely nontrivial and, in our opinion, realistic dynamics can be deduced. Nevertheless, we refer the readers to Chapter 10 for what we could call *extended CAR*, in which the number of levels can be greater than two.

Incidentally, we would like to notice that raising and lowering fermionic operators, as well as their related number operator, could be given in terms of Pauli matrices and spin operators. This particular representation of our operators, however, is not used in these notes. Again, we refer the readers to Chapter 10, and to Section 10.1.1 in particular, for further comments on Pauli matrices and their relation with Fermi operators.

This chapter is organized as follows. In Section 4.2, we consider a first simple model involving two populations and analyze the dynamics of their relationship starting from very natural assumptions. The considered model is linear and strictly local, and the equations of motion are solved analytically.

In Section 4.3, we extend this model by allowing a spatial distribution. The interaction is quadratic, so that the solution of the differential equations describing the time evolution can be again deduced analytically. The model is analyzed in terms of migrant and resident populations.

Section 4.4 contains some considerations on the same spatial model in the presence of a *delocalized* reservoir, used to mimic the existence of the outer world surrounding the two main populations.

In Section 4.5, we show how the same model introduced in Section 4.3, with a different choice of the parameters and of the initial conditions, can be used in the description of two competing populations.

Section 4.6 contains some final comments and a brief comparison with the existing literature.

## 4.2  A FIRST MODEL

We introduce here a first simple model, which is useful to fix the main ideas of our approach and the notation. This model will also be used as a building block for the more sophisticated model proposed in Sections 4.3–4.5. In particular, no spatial distribution is considered here, while the analysis of a two-dimensional distribution is our main interest in the next section. We associate to each population $S_j$ of the system $S$ an annihilation and a creation operator $a_j$ and $a_j^\dagger$, and a related number operator $\hat{n}_j :=$ $a_j^\dagger a_j$, whose mean value on a suitable state is identified here with the density of the population $S_j$. We consider two such populations, $S_1$ and $S_2$, and we assume the following anticommutation rules for $a_j$ and $a_j^\dagger$:

$$\{a_i, a_j^\dagger\} = \delta_{i,j} \,\mathbb{1}, \qquad \{a_i, a_j\} = \{a_i^\dagger, a_j^\dagger\} = 0, \qquad (4.1)$$

$i, j = 1, 2$. Here, as usual, $\mathbb{1}$ is the identity operator. Recall that $\{x, y\} :=$ $xy + yx$. These rules imply, in particular, that $a_j^2 = \left(a_j^\dagger\right)^2 = 0$. Hence, if $\varphi_{0,0}$ is the *ground vector* of $S$, $a_1\varphi_{0,0} = a_2\varphi_{0,0} = 0$, the only nontrivial, linearly independent, vectors of our Hilbert space $\mathcal{H}$ are

$$\varphi_{0,0}, \qquad \varphi_{1,0} := a_1^\dagger\varphi_{0,0}, \qquad \varphi_{0,1} := a_2^\dagger\varphi_{0,0}, \qquad \varphi_{1,1} := a_1^\dagger a_2^\dagger\varphi_{0,0}.$$

This means that $\dim(\mathcal{H}) = 4$. Translating in the present context the same ideas used in Chapter 3, we claim that the biological interpretation of these vectors can be deduced from the following eigenvalue equations:

$$\hat{n}_1 \varphi_{n_1,n_2} = n_1 \varphi_{n_1,n_2}, \qquad \hat{n}_2 \varphi_{n_1,n_2} = n_2 \varphi_{n_1,n_2}. \qquad (4.2)$$

Therefore, when we say that $\varphi_{0,0}$ is *the state of the system*, we mean that there are very few subjects of the two populations in our region. If the state of $\mathcal{S}$ is $\varphi_{1,0}$, there are very few elements of $\mathcal{S}_2$ but very many of $\mathcal{S}_1$. The opposite situation is described by $\varphi_{0,1}$, while $\varphi_{1,1}$ describes the case in which both populations coexist in the same area. As already stated, it is not possible to have, for example, more elements of $\mathcal{S}_1$ than those described by $\varphi_{1,0}$ or $\varphi_{1,1}$: trying to further increase the density of $\mathcal{S}_1$ simply destroys this population. From a mathematical point of view, this is a simple consequence of $(a_1^\dagger)^2 = 0$. From the biological side, this corresponds to the impossibility of putting too many members of a single species in a given constrained area. Of course, we could do better than this: in particular, we could consider a model with various intermediate values of the densities of $\mathcal{S}_1$ and $\mathcal{S}_2$. However, we do not consider this refinement here, because, as we have stressed earlier, these simplifying conditions give rise to an absolutely nontrivial dynamics.

The first step of our analysis consists in fixing a self-adjoint *Hamiltonian H* for the system $\mathcal{S}$, in order to deduce its dynamical behavior. Of course, $H$ is deduced mainly by fixing the form of the interaction between the populations. In particular, the operator assumed here, and written adopting **Rules 1** and **2** of Chapter 3, is the following:

$$H = H_0 + \lambda H_I, \qquad H_0 = \omega_1 a_1^\dagger a_1 + \omega_2 a_2^\dagger a_2, \qquad H_I = a_1^\dagger a_2 + a_2^\dagger a_1. \qquad (4.3)$$

Here, $\omega_j$ and $\lambda$ are real positive quantities, to ensure that $H$ is self-adjoint. In agreement with the quantum mechanical literature, here, $\omega_1$ and $\omega_2$ are called the *frequencies* of the populations. Of course, $\lambda$ must be taken to be 0 when the two populations do not interact. In this case, $H$ describes a somehow static situation, in which the densities of the two populations, described by the number operators $\hat{n}_j$, do not change with $t$. This is a consequence of the fact that $[H_0, \hat{n}_j] = 0$, $j = 1, 2$. Incidentally, this shows that **Rule 2** of Section 3.4 is respected by the Hamiltonian in Equation 4.3. We should notice, however, that the operators $a_j$ and $a_j^\dagger$ change with time. From this point of view, therefore, the system is not entirely stationary: even if the densities of the populations stay constant, some different operators (which, however, in general are not observables of the system) have a nontrivial time dependence. If $\lambda \neq 0$, $H$ contains also the contribution $H_I = a_1^\dagger a_2 + a_2^\dagger a_1$, which is written following **Rule 1** of Chapter 3, and can be understood as follows: $a_1^\dagger a_2$ makes the density of $S_1$ to increase (because of $a_1^\dagger$) and that of $S_2$ to decrease (because of $a_2$). The adjoint contribution, $a_2^\dagger a_1$, is responsible for the opposite phenomenon. The equations of motion are obtained using Equation 2.2:

$$\dot{a}_1(t) = -i\omega_1 a_1(t) - i\lambda a_2(t),$$
$$\dot{a}_2(t) = -i\omega_2 a_2(t) - i\lambda a_1(t), \qquad (4.4)$$

which can be solved with the initial conditions $a_1(0) = a_1$ and $a_2(0) = a_2$. The solution looks like

$$a_1(t) = \frac{1}{2\delta}\left(a_1\left((\omega_1 - \omega_2)\Phi_-(t) + \delta\Phi_+(t)\right) + 2\lambda a_2 \Phi_-(t)\right),$$
$$a_2(t) = \frac{1}{2\delta}\left(a_2\left(-(\omega_1 - \omega_2)\Phi_-(t) + \delta\Phi_+(t)\right) + 2\lambda a_1 \Phi_-(t)\right), \qquad (4.5)$$

where

$$\delta = \sqrt{(\omega_1 - \omega_2)^2 + 4\lambda^2},$$
$$\Phi_+(t) = 2\exp\left(-\frac{it(\omega_1 + \omega_2)}{2}\right)\cos\left(\frac{\delta t}{2}\right),$$
$$\Phi_-(t) = -2i\exp\left(-\frac{it(\omega_1 + \omega_2)}{2}\right)\sin\left(\frac{\delta t}{2}\right).$$

To fix the ideas, we assume here that $\delta \neq 0$. It is now easy to deduce the mean value of the time evolution of the number operator $\hat{n}_j(t)$, which, as discussed earlier, we interpret as *the density of $S_j$*: $n_j(t) := \langle \varphi_{n_1,n_2}, \hat{n}_j(t)\varphi_{n_1,n_2}\rangle$. More explicitly, $n_j(t)$ is the time evolution of the density of $S_j$ assuming that, at $t = 0$, the density of $S_1$ was $n_1$ and that of $S_2$ was $n_2$, the quantum numbers labeling the vector $\varphi_{n_1,n_2}$. Using the eigenvalue equation in (4.2), and the orthonormality of the different $\varphi_{n1,n2}$s, we obtain

$$n_1(t) = n_1\frac{(\omega_1 - \omega_2)^2}{(\omega_1 - \omega_2)^2 + 4\lambda^2} + \frac{4\lambda^2}{(\omega_1 - \omega_2)^2 + 4\lambda^2}$$
$$\times \left\{n_1\cos^2\left(\frac{\delta t}{2}\right) + n_2\sin^2\left(\frac{\delta t}{2}\right)\right\}, \qquad (4.6)$$

and

$$n_2(t) = n_2 \frac{(\omega_1 - \omega_2)^2}{(\omega_1 - \omega_2)^2 + 4\lambda^2} + \frac{4\lambda^2}{(\omega_1 - \omega_2)^2 + 4\lambda^2}$$
$$\times \left\{ n_2 \cos^2\left(\frac{\delta t}{2}\right) + n_1 \sin^2\left(\frac{\delta t}{2}\right) \right\}. \tag{4.7}$$

Notice that these formulas automatically imply that $n_1(t) + n_2(t) = n_1 + n_2$, independent of $t$ and $\lambda$. This is expected as $[H, \hat{n}_1 + \hat{n}_2] = 0$. Hence, the total density of the two species is preserved during the time evolution, even in the presence of interaction (i.e., when $\lambda \neq 0$). Second, as $n_1$ and $n_2$ can only be 0 or 1, it is easy to check that if $n_1 = n_2 = n$, $n = 0, 1$, then $n_1(t) = n_2(t) = n$ for all $t$: if the two populations have the same densities at $t = 0$, these remain the same also for $t > 0$. If, on the other hand, the densities are different, for instance, if $n_1 = 1$ and $n_2 = 0$ for $t = 0$, then

$$n_1(t) = 1 - \frac{4\lambda^2}{\delta^2}\sin^2\left(\frac{t\delta}{2}\right), \qquad n_2(t) = \frac{4\lambda^2}{\delta^2}\sin^2\left(\frac{t\delta}{2}\right),$$

while if $n_1 = 0$ and $n_2 = 1$, then

$$n_2(t) = 1 - \frac{4\lambda^2}{\delta^2}\sin^2\left(\frac{t\delta}{2}\right), \qquad n_1(t) = \frac{4\lambda^2}{\delta^2}\sin^2\left(\frac{t\delta}{2}\right).$$

In all these cases, we have $0 \leq n_j(t) \leq 1$ for all $t$, as it should be.

As $n_1(t) + n_2(t) = n_1 + n_2$, we now find that $\delta n(t) := |n_1(t) - n_1| = |n_2(t) - n_2|$, which, in the two cases above, gives $\delta n(t) = \frac{4\lambda^2}{\delta^2}\sin^2\left(\frac{t\delta}{2}\right)$; in particular, therefore, the absolute value of the variations of the two populations coincide. In general, Equation 4.6 gives $n_1(t) - n_1 = \frac{4\lambda^2}{\delta^2}(n_2 - n_1)\sin^2\left(\frac{t\delta}{2}\right)$, which is in agreement with the previous result, recalling that $n_1$ and $n_2$ can only be 0 or 1. Note also that, if in particular, $n_1 = n_2$, then $n_1(t) = n_1$ (and $n_2(t) = n_2$) for all $t$, as we have deduced previously.

Let us now restrict, to be concrete, to the case $n_1 = 1$ and $n_2 = 0$. Hence $\Delta_n := \max\{\delta n(t)\} = \frac{1}{1+(\omega_1-\omega_2/2\lambda)^2}$, for $\frac{t\delta}{2} = \frac{\pi}{2}, \frac{3\pi}{2}, \ldots$, while $\min\{\delta n(t)\} = 0$ for $\frac{t\delta}{2} = 0, \pi, 2\pi, \ldots$. In particular, $\Delta_n$ is almost equal to 1 if $\omega_1 \simeq \omega_2$, independently of $\lambda \neq 0$, while is almost 0, as $\lambda$ is kept fixed, when $|\omega_1 - \omega_2|$ becomes very large. Incidentally, if $\lambda = 0$, the two species do not interact and, in fact, $\Delta_n = 0$: the observables of the model have essentially no dynamics. For obvious reasons, we are mainly interested in $\lambda > 0$. As already mentioned, these results show that the free Hamiltonian, which does not affect the density of the populations if $\lambda = 0$, produces a nontrivial effect on the dynamics of the populations when the interaction is not 0, $\lambda \neq 0$. Similar conclusions are deduced in Sections 4.3 and 4.5, while considering the spatial version of this model. More in details, $|\omega_1 - \omega_2|$ can be considered as a sort of *inertia* of the system: the larger its value, the smaller are the variations of $n_j(t) - n_j$. On the other hand, if $|\omega_1 - \omega_2| \simeq 0$, then the system has almost no inertia and, in fact, very large changes in the densities of both the populations occur as $\Delta_n \simeq 1$. It is interesting to remark that only the difference between the two frequencies $\omega_1$ and $\omega_2$ plays a role in the dynamics of both $\mathcal{S}_1$ and $\mathcal{S}_2$.

Similar conclusions can also be deduced from Equations 4.6 and 4.7: if $|\omega_1 - \omega_2| \gg 2\lambda$ there is essentially no dynamics as $n_j(t) \simeq n_j$, $j = 1, 2$. On the contrary, if $|\omega_1 - \omega_2| \ll 2\lambda$, the constant contributions in Equations 4.6 and 4.7 are very small compared with the oscillating contributions.

## 4.3 A SPATIAL MODEL

In this section, we extend the model considered earlier, including also spatial effects: we consider a two-dimensional region $\mathcal{R}$ in which, in principle, the two populations are distributed. Under reasonable assumptions, a simple model for $\mathcal{S}_1$ and $\mathcal{S}_2$ is now constructed, and its dynamics investigated.

The starting point is the rectangular (or square) region $\mathcal{R}$, which we divide into $N$ cells, labeled by $\alpha = 1, 2, \ldots, N = L \cdot L'$. With $\alpha - 1$, we label the first cell down to the left, while $N$ labels the last cell, up to the right (Figure 4.1).

The main idea to build up our spatial model is that, in each cell $\alpha$, the two populations, whose related operators are $a_\alpha$, $a_\alpha^\dagger$, and $\hat{n}_\alpha^{(a)} = a_\alpha^\dagger a_\alpha$ for what concerns $\mathcal{S}_1$, and $b_\alpha$, $b_\alpha^\dagger$, and $\hat{n}_\alpha^{(b)} = b_\alpha^\dagger b_\alpha$ for $\mathcal{S}_2$, behave as in Section 4.2. This means that the same Hamiltonian as in Equation 4.3 is assumed here in each cell $\alpha$. Therefore, using our new simplifying notation, we define

$$H_\alpha = H_\alpha^0 + \lambda_\alpha H_\alpha^I, \qquad H_\alpha^0 = \omega_\alpha^a a_\alpha^\dagger a_\alpha + \omega_\alpha^b b_\alpha^\dagger b_\alpha, \qquad H_\alpha^I = a_\alpha^\dagger b_\alpha + b_\alpha^\dagger a_\alpha. \tag{4.8}$$

Extending what we have discussed in the previous section, it is natural to interpret the operators $\hat{n}_\alpha^{(a)}$ and $\hat{n}_\alpha^{(b)}$ as *local density operators* (the local densities are in the sense of mixtures; hence, we may sum up the local densities relative to different cells) of the two populations in the cell $\alpha$: if the mean value of, say, $\hat{n}_\alpha^{(a)}$, in the state of the system is equal to

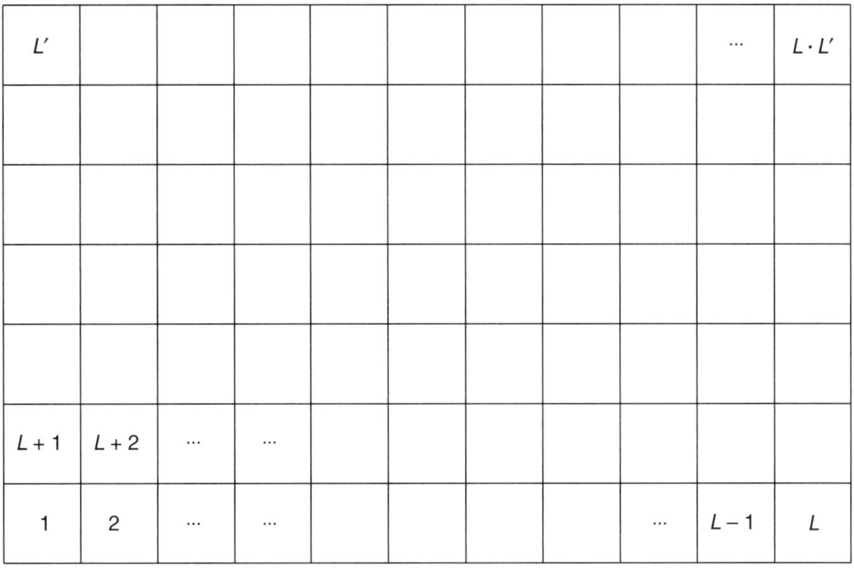

| $L'$ |  |  |  |  |  |  |  |  | $\cdots$ | $L \cdot L'$ |
|---|---|---|---|---|---|---|---|---|---|---|
|  |  |  |  |  |  |  |  |  |  |  |
|  |  |  |  |  |  |  |  |  |  |  |
|  |  |  |  |  |  |  |  |  |  |  |
|  |  |  |  |  |  |  |  |  |  |  |
| $L+1$ | $L+2$ | $\cdots$ | $\cdots$ |  |  |  |  |  |  |  |
| 1 | 2 | $\cdots$ | $\cdots$ |  |  |  |  | $\cdots$ | $L-1$ | $L$ |

**Figure 4.1**    The two-dimensional lattice for the spatial model.

one, this means that the density of $\mathcal{S}_1$ in the cell $\alpha$ is very high. Note that $H_\alpha = H_\alpha^\dagger$ as all the parameters, which in general are assumed to be cell dependent (to allow for the description of an anisotropic situation), are real and positive numbers. The anticommutation rules extend those in Equation 4.1:

$$\{a_\alpha, a_\beta^\dagger\} = \{b_\alpha, b_\beta^\dagger\} = \delta_{\alpha,\beta} \, \mathbb{1},\qquad (4.9)$$

whereas all the other anticommutators are trivial: $\{a_\alpha, b_\beta\} = \{a_\alpha, b_\beta^\dagger\} = \{a_\alpha, a_\beta\} = \{b_\alpha, b_\beta\} = 0$. Of course, the Hamiltonian $H$ for the full system must consist of a sum of all the different $H_\alpha$ plus another contribution, $H_{\text{diff}}$, responsible for the diffusion of the populations all around the lattice. A natural choice for $H_{\text{diff}}$ is the following one:

$$H_{\text{diff}} = \sum_{\alpha,\beta} p_{\alpha,\beta} \left\{ \gamma_a \left( a_\alpha a_\beta^\dagger + a_\beta a_\alpha^\dagger \right) + \gamma_b \left( b_\alpha b_\beta^\dagger + b_\beta b_\alpha^\dagger \right) \right\}, \qquad (4.10)$$

where $\gamma_a$, $\gamma_b$, and the $p_{\alpha,\beta}$ are real quantities, to keep $H_{\text{diff}}$ self-adjoint. In particular, $p_{\alpha,\beta}$ can only be 0 or 1 depending on the possibility of the populations to move from cell $\alpha$ to cell $\beta$, or vice versa. In fact, if $p_{\alpha,\beta} = 1$, the term $a_\alpha a_\beta^\dagger$ produces a lowering in the density of $\mathcal{S}_1$ in the cell $\alpha$ (because of $a_\alpha$), with a related raising of the density of the same species in $\beta$ (because of $a_\beta^\dagger$): some member of $\mathcal{S}_1$ is moving from $\alpha$ to $\beta$. For this reason, the $p_{\alpha,\beta}$'s can be considered as *diffusion coefficients*, more or less like $\gamma_a$ and $\gamma_b$: if, for instance, $\gamma_a = 0$ and $\gamma_b > 0$, it is clear that no diffusion is possible for $S_1$, while $S_2$ can move from one cell to another all along $\mathcal{R}$. However, we should stress that, also when $\gamma_a = 0$, there exists still the possibility of increasing, after some time, the density of $S_1$ in a part of $\mathcal{R}$ where this density was initially very small. This is because the self-adjoint Hamiltonian $H = \sum_\alpha H_\alpha + H_{\text{diff}}$ also contains a contribution $H_\alpha^I = a_\alpha^\dagger b_\alpha + b_\alpha^\dagger a_\alpha$, which allows the density of $S_1$ to increase just as a consequence of its interaction with $S_2$ in the cell $\alpha$. This is not a diffusion but a *competition* effect, which is localized in a single cell of $\mathcal{R}$!

In the rest of this chapter, we assume that diffusion may take place only between nearest neighboring cells. Of course, we should clarify what we mean by *neighboring*: here, we consider a simple planar topology, in which the neighboring cells of the cell labeled $\alpha$ are the cells labeled $\alpha - 1$, $\alpha + 1$, $\alpha + L$, and $\alpha - L$, provided that they exist (this is verified for the internal cells of the lattice); for the cells located along the boundaries of the lattice we have only three neighbors (two neighbors for the four cells located at the corners of the lattice). The situation is described in Figure 4.2, where the arrows show the possible movements of the members of a given species occupying that particular cell of $\mathcal{R}$. From a practical point of view, this means, for instance, that $p_{1,1} = p_{1,L+1} = 1$, while $p_{1,j} = 0$ for all $j \neq 1, L + 1$, and so on.

This is a natural choice for the physical system we have in mind here. However, different choices could also be of some interest. For instance, we could also use a torus topology in which all the cells have four neighboring cells. To deal with these different topologies, it would be enough to modify the values of the diffusion coefficients, but we do not consider these possibilities here. We also assume that $p_{\alpha,\alpha} = 0$ and that $p_{\alpha,\beta} = p_{\beta,\alpha}$: this last condition implies that, if $S_1$ can move from $\alpha$ to $\beta$, then it can also return, moving in the opposite direction. The differential equations for the annihilation operators, obtained by Equation 2.2 using the Hamiltonian $H$, read as follows

$$\begin{cases} \dot{a}_\alpha = -i\omega_\alpha^a a_\alpha - i\lambda_\alpha b_\alpha + 2i\gamma_a \sum_\beta p_{\alpha,\beta} a_\beta, \\ \dot{b}_\alpha = -i\omega_\alpha^b b_\alpha - i\lambda_\alpha a_\alpha + 2i\gamma_b \sum_\beta p_{\alpha,\beta} b_\beta, \end{cases} \tag{4.11}$$

which are linear and, therefore, not particularly difficult to solve, in principle.

### 4.3.1 A Simple Case: Equal Coefficients

As a first step, we suppose here that $\omega_\alpha^b = \omega_\alpha^a = \omega$, $\lambda_\alpha = \lambda$, and $\gamma_a = \gamma_b = \tilde{\gamma}$, for all possible $\alpha$. Hence, by introducing $a_\alpha(t) = A_\alpha(t)e^{-i\omega t}$ and

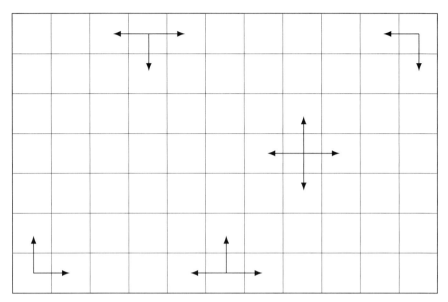

**Figure 4.2**    Possible movements in $\mathcal{R}$.

$b_\alpha(t) = B_\alpha(t)e^{-i\omega t}$, the earlier equations can be rewritten as

$$\begin{cases} \dot{A}_\alpha = -i\lambda B_\alpha + 2i\tilde{\gamma} \sum_\beta p_{\alpha,\beta} A_\beta, \\ \dot{B}_\alpha = -i\lambda B_\alpha + 2i\tilde{\gamma} \sum_\beta p_{\alpha,\beta} B_\beta. \end{cases} \tag{4.12}$$

Note that these equations do not depend on the size of the region $\mathcal{R}$.

The diffusion coefficients $p_{\alpha,\beta}$ will all be 0 except when $\alpha$ and $\beta$ refer to nearest neighbors in the planar topology; see Figure 4.2. In this case $p_{\alpha,\beta} = 1$. We consider now the situation of a square region $\mathcal{R}$ with $N = L^2$ cells, starting with the simplest nontrivial situation, $L = 2$. In this case, the only nonzero diffusion coefficients are $p_{1,2}$, $p_{1,3}$, $p_{2,1}$, $p_{2,4}$, $p_{3,1}$, $p_{3,4}$, $p_{4,2}$, and $p_{4,3}$, which are all equal to 1, while all the others are 0. Fixing *natural* initial conditions for the annihilation operators, $A_\alpha(0) = a_\alpha$ and $B_\alpha(0) = b_\alpha$, we get, for instance,

$$\begin{aligned} A_1(t) = \frac{1}{4}[&2(a_1 - a_4)\cos(t) + (a_1 + a_4 + b_2 + b_3)\cos(7t) \\ &+ (a_1 + a_4 - b_2 - b_3)\cos(9t) \\ &- i(2(b_1 - b_4 + (b_1 + b_4)\cos(8t))\sin(t) \\ &- (a_2 + a_3)(\sin(7t) + \sin(9t)))], \end{aligned}$$

and so on. The number operators (i.e., the local densities of $S_1$ and $S_2$) are deduced directly from the capital operators $A_\gamma(t)$ and $B_\gamma(t)$ as $\hat{n}_\gamma^{(a)}(t) := a_\gamma^\dagger(t)a_\gamma(t) = A_\gamma^\dagger(t)A_\gamma(t)$ and $\hat{n}_\gamma^{(b)}(t) := b_\gamma^\dagger(t)b_\gamma(t) = B_\gamma^\dagger(t)B_\gamma(t)$. Assuming that for $t = 0$ both the populations are concentrated in the cell $1^1$, $n_1^{(a)}(0) = n_1^{(b)}(0) = 1$, while $n_\alpha^{(a)}(0) = n_\alpha^{(b)}(0) = 0$ for $\alpha = 2, 3, 4$, we find

$$\begin{cases} n_1^{(a)}(t) = n_1^{(b)}(t) = (\cos(4t))^4, \\ n_2^{(a)}(t) = n_2^{(b)}(t) = \dfrac{1}{4}(\sin(8t))^2, \\ n_3^{(a)}(t) = n_3^{(b)}(t) = \dfrac{1}{4}(\sin(8t))^2, \\ n_4^{(a)}(t) = n_4^{(b)}(t) = (\sin(4t))^4. \end{cases} \tag{4.13}$$

These results are deduced by taking the mean values of the operators $\hat{n}_\gamma^{(a)}(t)$ and $\hat{n}_\gamma^{(b)}(t)$ on a vector state $\omega_{\mathbf{n}^a,\mathbf{n}^b}(.) = \langle \varphi_{\mathbf{n}^a,\mathbf{n}^b}, \cdot \varphi_{\mathbf{n}^a,\mathbf{n}^b}\rangle$, in which $\mathbf{n}^a = (1, 0, 0, 0)$ and $\mathbf{n}^b = (1, 0, 0, 0)$.

The results in Equation 4.13 look quite reasonable. In fact,

1. recalling that, because of our simplifying assumptions, all the parameters of $a$ and $b$ coincide, and as the initial conditions for $n_j^{(a)}(t)$ and $n_j^{(b)}(t)$ coincide as well, it is clear that the time behavior of the two populations must be identical;

2. at $t = 0$ only the cell 1 is populated. Hence, the initial conditions for the two local densities are respected;

3. we observe that $n_2^{(a)}(t) = n_3^{(a)}(t)$ and $n_2^{(b)}(t) = n_3^{(b)}(t)$; this is not surprising as, because of the isotropy of $\mathcal{R}$, there is an equal probability for, say, a member of $S_1$ to move from cell 1 to cell 2 or to cell 3; so he can reach cell 4 only through cells 2 or 3, but not directly;

4. this explains why, plotting $n_1^{(a)}(t)$, $n_2^{(a)}(t) = n_3^{(a)}(t)$, and $n_4^{(a)}(t)$, for small values of $t$, $n_2^{(a)}(t)$ increases faster than $n_4^{(a)}(t)$ but, after some time, $n_4^{(a)}(t)$ becomes larger than, say, $n_2^{(a)}(t)$; this is because after cells 2 and 3 are populated, they both start contributing to the population of cell 4. This result is shown in Figure 4.3.

---

[1] Of course, this choice is not very relevant in the context of migration but is useful here just to fix the ideas. The application to migration is considered for larger lattice, where the situation is surely more interesting and a bit more realistic.

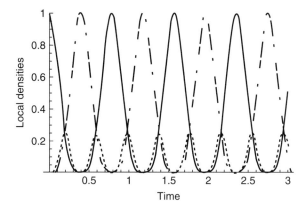

**Figure 4.3** $n_1^{(a)}(t)$, solid line, $n_2^{(a)}(t) = n_3^{(a)}(t)$, dotted line, and $n_4^{(a)}(t)$, dotted-dashed line.

Let us now consider a slightly larger lattice, with $L = 3$. In this case, $\mathcal{R}$ is made up of nine cells and the only nonzero diffusion coefficients are (mentioning just one between $p_{\alpha,\beta}$ and $p_{\beta,\alpha}$) $p_{1,2}$, $p_{1,4}$, $p_{2,5}$, $p_{2,3}$, $p_{3,6}$, $p_{4,5}$, $p_{4,7}$, $p_{5,6}$, $p_{5,8}$, $p_{6,9}$, $p_{7,8}$, and $p_{8,9}$, which are all equal to 1. The 18 differential equations deduced from those in Equation 4.12 can be written as

$$\dot{X}_9 = i\mathcal{M}_9 X_9, \qquad \mathcal{M}_9 = 2\tilde{\gamma} M_9 - \lambda J_9, \tag{4.14}$$

where we have introduced the following vector and matrices:

$$X_9 = \begin{pmatrix} A_1 \\ A_2 \\ \cdots \\ \cdots \\ A_9 \\ B_1 \\ B_2 \\ \cdots \\ \cdots \\ B_9 \end{pmatrix}, \quad N_9 = \begin{pmatrix} 0 & 1 & 0 & 1 & 0 & 0 & 0 & 0 & 0 \\ 1 & 0 & 1 & 0 & 1 & 0 & 0 & 0 & 0 \\ 0 & 1 & 0 & 0 & 0 & 1 & 0 & 0 & 0 \\ 1 & 0 & 0 & 0 & 1 & 0 & 1 & 0 & 0 \\ 0 & 1 & 0 & 1 & 0 & 1 & 0 & 1 & 0 \\ 0 & 0 & 1 & 0 & 1 & 0 & 0 & 0 & 1 \\ 0 & 0 & 0 & 1 & 0 & 0 & 0 & 1 & 0 \\ 0 & 0 & 0 & 0 & 1 & 0 & 1 & 0 & 1 \\ 0 & 0 & 0 & 0 & 0 & 1 & 0 & 1 & 0 \end{pmatrix},$$

$$M_9 = \begin{pmatrix} N_9 & \mathbf{0}_9 \\ \mathbf{0}_9 & N_9 \end{pmatrix}, \quad J_9 = \begin{pmatrix} \mathbf{0}_9 & \mathbb{1}_9 \\ \mathbb{1}_9 & \mathbf{0}_9 \end{pmatrix},$$

where, with the same notation of Chapter 3, $\mathbf{0}_9$ and $\mathbb{1}_9$ are, respectively, the $9 \times 9$ zero and identity matrices. Note that $\mathcal{M}_9$ is a symmetric real matrix.

The generalization to larger $\mathcal{R}$ is straightforward. In this case, we have

$$\dot{X}_{L^2} = i\mathcal{M}_{L^2}X_{L^2}, \qquad \mathcal{M}_{L^2} = 2\tilde{\gamma}M_{L^2} - \lambda J_{L^2}. \tag{4.15}$$

Here, the transpose of $X_{L^2}$ is $(A_1, A_2, \ldots, A_{L^2}, B_1, B_2, \ldots, B_{L^2})$, $\mathbf{0}_{L^2}$, $\mathbb{1}_{L^2}$ and $J_{L^2}$ extend those given earlier, and

$$M_{L^2} = \begin{pmatrix} N_{L^2} & \mathbf{0}_{L^2} \\ \mathbf{0}_{L^2} & N_{L^2} \end{pmatrix}.$$

Once again, $\mathcal{M}_{L^2}$ is a symmetric real matrix. Of course, the explicit form of the matrix $N_{L^2}$ can be constructed extending the previous considerations: this matrix has all zero entries except those matrix elements that correspond to nearest neighbors cells, all equal to one.

The solution of Equation 4.15 is, formally, very simple (because of the linearity of the differential equations):

$$X_{L^2}(t) = \exp\left(i\,\mathcal{M}_{L^2}t\right)X_{L^2}(0). \tag{4.16}$$

Let us call $d_{\alpha,\beta}(t)$ the generic entry of the matrix $\exp(i\,\mathcal{M}_{L^2}t)$, and let us assume that at $t = 0$ the system is described by the vector $\varphi_{\mathbf{n}^a,\mathbf{n}^b}$, where $\mathbf{n}^a = (n_1^a, n_2^a, \ldots, n_{L^2}^a)$ and $\mathbf{n}^b = (n_1^b, n_2^b, \ldots, n_{L^2}^b)$. Hence, the mean values of the time evolution of the number operators in the cell $\alpha$,

$$n_\alpha^a(t) := \langle \varphi_{\mathbf{n}^a,\mathbf{n}^b}, a_\alpha^\dagger(t)a_\alpha(t)\varphi_{\mathbf{n}^a,\mathbf{n}^b} \rangle = \langle \varphi_{\mathbf{n}^a,\mathbf{n}^b}, A_\alpha^\dagger(t)A_\alpha(t)\varphi_{\mathbf{n}^a,\mathbf{n}^b} \rangle,$$

$$n_\alpha^b(t) := \langle \varphi_{\mathbf{n}^a,\mathbf{n}^b}, b_\alpha^\dagger(t)b_\alpha(t)\varphi_{\mathbf{n}^a,\mathbf{n}^b} \rangle = \langle \varphi_{\mathbf{n}^a,\mathbf{n}^b}, B_\alpha^\dagger(t)B_\alpha(t)\varphi_{\mathbf{n}^a,\mathbf{n}^b} \rangle,$$

can be written as

$$\begin{cases} n_\alpha^a(t) = \sum_{\theta=1}^{L^2} |d_{\alpha,\theta}(t)|^2 n_\theta^a + \sum_{\theta=1}^{L^2} |d_{\alpha,L^2+\theta}(t)|^2 n_\theta^b, \\ n_\alpha^b(t) = \sum_{\theta=1}^{L^2} |d_{L^2+\alpha,\theta}(t)|^2 n_\theta^a + \sum_{\theta=1}^{L^2} |d_{L^2+\alpha,L^2+\theta}(t)|^2 n_\theta^b. \end{cases} \tag{4.17}$$

These formulas describe the densities of $\mathcal{S}_1$ and $\mathcal{S}_2$ in all the cells of the region $\mathcal{R}$.

### 4.3.2 Back to the General Case: Migration

The same strategy, which produces the solution in Equation 4.17, can be used to solve system 4.11. In this case, Equation 4.15 is replaced by a similar equation,

$$\dot{X}_{L^2} = i\mathcal{K}_{L^2}X_{L^2}, \tag{4.18}$$

where $\mathcal{K}_{L^2} = 2T_{L^2} - P_{L^2}$, with $T_{L^2}$ and $P_{L^2}$ two $2L^2 \times 2L^2$ matrices defined as follows:

$$T_{L^2} = \begin{pmatrix} N_{L^2} \text{ with } 1 \text{ replaced by } \gamma_a & 0 \\ 0 & N_{L^2} \text{ with } 1 \text{ replaced by } \gamma_b \end{pmatrix},$$

and

$$P_{L^2} = \begin{pmatrix} \Omega^{(a)} & \Lambda \\ \Lambda & \Omega^{(b)} \end{pmatrix}.$$

In $P_{L^2}$ we have introduced the following matrices: $\Omega^{(a)} = \text{diag}\{\omega_1^a, \omega_2^a, \dots, \omega_{L^2}^a\}$, $\Omega^{(b)} = \text{diag}\{\omega_1^b, \omega_2^b, \dots, \omega_{L^2}^b\}$, and $\Lambda = \text{diag}\{\lambda_1, \lambda_2, \dots, \lambda_{L^2}\}$.

The solution of Equation 4.18 is analogous to that in Equation 4.16:

$$X_{L^2}(t) = \exp\left(i\mathcal{K}_{L^2}t\right) X_{L^2}(0).$$

Calling $f_{\alpha,\beta}(t)$ the generic entry of the matrix $\exp\left(i\,\mathcal{K}_{L^2}t\right)$, and repeating the same procedure as earlier, we get

$$\begin{cases} n_\alpha^a(t) = \sum_{\theta=1}^{L^2} |f_{\alpha,\theta}(t)|^2\, n_\theta^a + \sum_{\theta=1}^{L^2} |f_{\alpha,L^2+\theta}(t)|^2\, n_\theta^b, \\ n_\alpha^b(t) = \sum_{\theta=1}^{L^2} |f_{L^2+\alpha,\theta}(t)|^2\, n_\theta^a + \sum_{\theta=1}^{L^2} |f_{L^2+\alpha,L^2+\theta}(t)|^2\, n_\theta^b, \end{cases} \tag{4.19}$$

which generalize the result in Equation 4.17.

These formulas are now used to deduce the local densities of the two populations $S_1$ and $S_2$ in three different regions. The first region, $\mathcal{R}_1$, corresponding to cells $1, 2, L+1$, and $L+2$ (bottom-left corner of $\mathcal{R}$), is that part of $\mathcal{R}$ where all the members of $S_1$ are originally (i.e., at $t = 0$) localized. Population $S_2$, at $t = 0$, is assumed to be localized in the four cells $L^2 - L - 1$, $L^2 - L$, $L^2 - 1$, and $L^2$, our region $\mathcal{R}_2$ (top-right corner of $\mathcal{R}$). All the other cells belong to $\mathcal{R}_3$, that part of $\mathcal{R}$, which must be crossed by the populations to move from $\mathcal{R}_1$ to $\mathcal{R}_2$ or vice versa. Just to fix the ideas, we could think of $S_1$ and $S_2$ as people from Africa ($R_1$) and Europe ($R_2$), respectively, and the Mediterranean sea as the region $R_3$. Alternatively, going back in time, we could think of $R_1$ as South Italy, $R_2$ as North America, and $R_3$ as the Atlantic Ocean[2]. In our numerical computations, we have fixed $L = 11$. This choice produces a lattice that is not too small to be trivial and not too large to generate serious numerical difficulties. In Figures 4.4–4.9, we plot the two *local densities* (the sum of the densities in the different cells) for $S_1$ (solid line) and $S_2$ (dashed line), in $R_1$ (panels a), $R_2$ (panels b), and $R_3$ (panels c), for different choices of the parameters and for the initial conditions considered above.

In particular, in Figures 4.4–4.6, the parameters $\lambda_\alpha$s are taken to be equal everywhere: $\lambda_\alpha = 0.05$, in all the cells. On the other hand, in Figures 4.7–4.9, $\lambda_\alpha = 0.05$ in $R_3$, while $\lambda_\alpha = 0.2$ in $R_1$ and $R_2$. This difference is useful to model the fact that $S_1$ and $S_2$ most probably interact where they live, rather than *on the way*.

All these figures share a common feature: they all show that $S_1$ leaves $R_1$, moving toward $R_2$, while only a small part of $S_2$ moves toward $R_1$. This is related to the value of the parameters $\gamma_a$ and $\gamma_b$, as well as to the $p_{\alpha,\beta}$ that were fixed at the very beginning, accounting for the diffusion in the model; see Equation 4.10. As $\gamma_a > \gamma_b$, it is clear that $S_1$ has a larger mobility than $S_2$ or, in other terms, that $S_1$ tends to leave Africa much more than $S_2$ to leave Europe: $\gamma_a$ and $\gamma_b$ are the *strength of migration* of the two populations. This is exactly the content of all the figures. Moreover, Figures 4.4, 4.5, plots (b), show that, when the density of $S_1$ in $R_2$ approaches that of $S_2$, $S_2$ reacts very strongly in two ways: their birth rate increases very fast (in fact, its density increases) and they start rejecting somehow the members of $S_1$ (and indeed, the density of $S_1$ suddenly decreases). After this first reaction, we see that, from time to time, a certain amount of people of $S_1$ go back to $R_1$. In our settings, this can be interpreted as if these people had reached a certain welfare. We also see that in $R_2$ the density of $S_2$ stays almost always larger than that of $S_1$, while in $R_1$ the density of $S_2$ is always very low: rich people do not like much to go to a poor area! Moreover, a lot of people of both populations are in $R_3$: they travel, not necessarily moving from $R_1$ to $R_2$, or vice versa. The plots also suggest that the $\omega$'s parameters measure a sort of *inertia* of the two populations: increasing the value of, say, $\omega_b$, reduces the oscillatory behavior of $S_2$, as we can see from Figures 4.4–4.6. Analogously, we have checked that increasing the value of $\omega_a$, we get a more *static* behavior for $S_1$. This conclusion is similar to what we

---

[2]This is, let me say, a *familiar* choice: some of my ancestors moved, at the beginning of the twentieth century, from Sicily to North America.

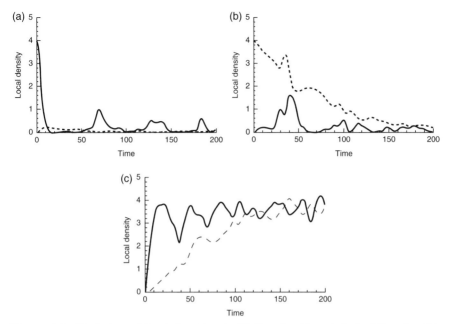

**Figure 4.4**  Evolution of local densities (solid line for $\mathcal{S}_1$ and dashed line for $\mathcal{S}_2$). (a) Africa; (b) Europe; and (c) Mediterranean Sea. $\gamma_a = 0.1$, $\gamma_b = 0.004$, $\omega_\alpha^a = 1$, $\omega_\alpha^b = 0.3$, $\lambda_\alpha = 0.05$, $\forall \alpha \in \mathcal{R}$.

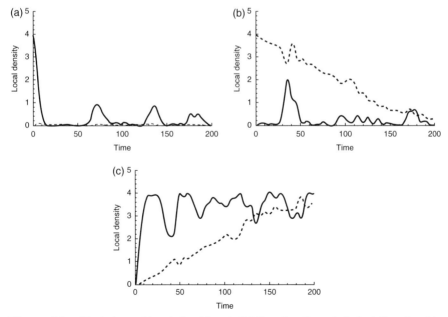

**Figure 4.5**  Evolution of local densities (solid line for $\mathcal{S}_1$ and dashed line for $\mathcal{S}_2$). (a) Africa; (b) Europe; and (c) Mediterranean Sea. $\gamma_a = 0.1$, $\gamma_b = 0.004$, $\omega_\alpha^a = 1$, $\omega_\alpha^b = 1$, $\lambda_\alpha = 0.05$, $\forall \alpha \in \mathcal{R}$.

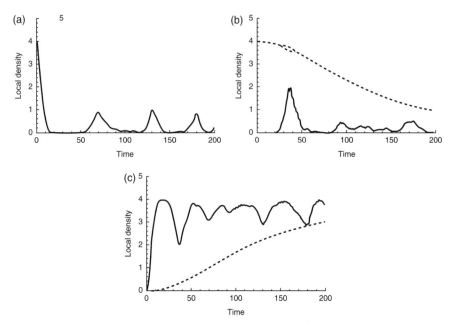

**Figure 4.6**    Evolution of local densities (solid line for $S_1$ and dashed line for $S_2$). (a) Africa; (b) Europe; and (c) Mediterranean Sea. $\gamma_a = 0.1$, $\gamma_b = 0.004$, $\omega_\alpha^a = 1$, $\omega_\alpha^b = 3$, $\lambda_\alpha = 0.05$, $\forall \alpha \in \mathcal{R}$.

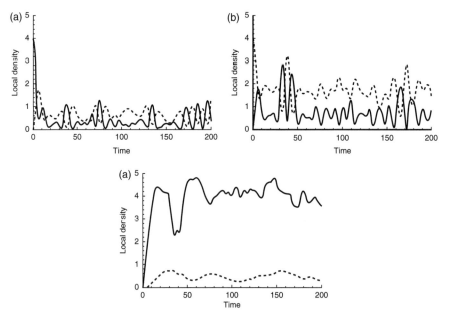

**Figure 4.7**    Evolution of local densities (solid line for $S_1$ and dashed line for $S_2$). (a) Africa; (b) Europe; and (c) Mediterranean Sea. $\gamma_a = 0.1$, $\gamma_b = 0.004$, $\omega_\alpha^a = 1$, $\omega_\alpha^b = 0.3$, $\forall \alpha \in \mathcal{R}$; $\lambda_\alpha = 0.2$ for $\alpha \in \mathcal{R}_1 \cup \mathcal{R}_2$ and $\lambda_\alpha = 0.05$ for $\alpha \in \mathcal{R}_3$.

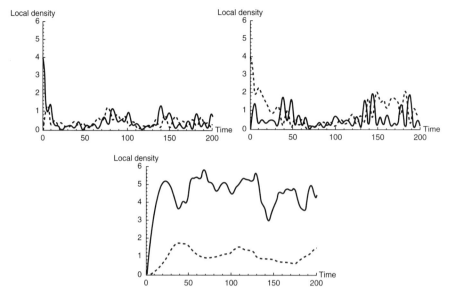

**Figure 4.8**  Evolution of local densities (solid line for $S_1$ and dashed line for $S_2$). (a) Africa; (b) Europe; (c) Mediterranean Sea. $\gamma_a = 0.1$, $\gamma_b = 0.004$, $\omega_\alpha^a = 1$, $\omega_\alpha^b = 1$ $\forall \alpha \in \mathcal{R}$; $\lambda_\alpha = 0.2$ for $\alpha \in \mathcal{R}_1 \cup \mathcal{R}_2$ and $\lambda_\alpha = 0.05$ for $\alpha \in \mathcal{R}_3$.

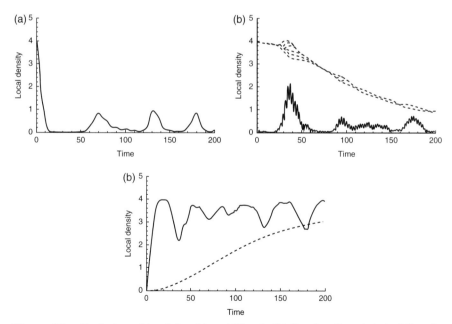

**Figure 4.9**  Evolution of local densities. Africa (solid line for $S_1$ and dashed line for $S_2$). (a) Africa; (b) Europe; and (c) Mediterranean Sea. $\gamma_a = 0.1$, $\gamma_b = 0.004$, $\omega_\alpha^a = 1$, $\omega_\alpha^b = 2$ $\forall \alpha \in \mathcal{R}$; $\lambda_\alpha = 0.2$ for $\alpha \in \mathcal{R}_1 \cup \mathcal{R}_2$ and $\lambda_\alpha = 0.05$ for $\alpha \in \mathcal{R}_3$.

have deduced in Section 4.2. We should also mention that our numerical computations for $L > 11$ confirm our conclusions, showing that the size of $\mathcal{R}$ does not really affect our understanding of the system, but only modifies the time needed to move from one corner of the lattice to the other. Figures 4.7–4.9 show much faster oscillations in the densities of $\mathcal{S}_1$ and $\mathcal{S}_2$ than those in Figures 4.4–4.6, in particular, in the regions $\mathcal{R}_1$ and $\mathcal{R}_2$. This is due to the fact that, in these regions, the interaction parameter between the populations $\lambda_\alpha$ is taken larger than before. Hence, their densities can change faster: we are in a regime in which the interaction between $\mathcal{S}_1$ and $\mathcal{S}_2$ becomes more important than the diffusion of the populations! This conclusion is in agreement with the expression of $H$ and with the meaning of its various contributions.

## 4.4   THE ROLE OF A RESERVOIR

It may be interesting to see, at least at a qualitative level, what changes in the model considered earlier when the two populations $\{a_\alpha\}$ and $\{b_\alpha\}$, other than interacting between themselves and diffusing in the region $\mathcal{R}$, interact with a reservoir $\tilde{R}$ described by two sets of fermionic operators, $\{A_\alpha(k), \, k \in \mathbb{R}\}$ and $\{B_\alpha(k), \, k \in \mathbb{R}\}$, in a way that is very close to that discussed in Section 3.4. More in details, we define the following Hamiltonian

$$H = H_{a,A} + H_{b,B} + H_{a,b} + H_{\text{diff}}, \tag{4.20}$$

where, as in the previous section

$$H_{a,b} = \lambda \sum_\alpha \left( a_\alpha^\dagger b_\alpha + b_\alpha^\dagger a_\alpha \right)$$

and

$$H_{\text{diff}} = \sum_{\alpha,\beta} p_{\alpha,\beta} \left\{ \gamma_a \left( a_\alpha a_\beta^\dagger + a_\beta a_\alpha^\dagger \right) + \gamma_b \left( b_\alpha b_\beta^\dagger + b_\beta b_\alpha^\dagger \right) \right\},$$

whereas

$$H_{a,A} = \sum_\alpha \left[ \omega_\alpha^a a_\alpha^\dagger a_\alpha + \int_{\mathbb{R}} \Omega_\alpha^A(k) \, A_\alpha^\dagger(k) A_\alpha(k) \, dk \right.$$
$$\left. + \mu_A \int_{\mathbb{R}} \left( a_\alpha^\dagger A_\alpha(k) + A_\alpha^\dagger(k) a_\alpha \right) dk \right]$$

and

$$H_{b,B} = \sum_{\alpha} \left[ \omega_{\alpha}^{b} b_{\alpha}^{\dagger} b_{\alpha} + \int_{\mathbb{R}} \Omega_{\alpha}^{B}(k)\, B_{\alpha}^{\dagger}(k) B_{\alpha}(k)\, dk \right.$$
$$\left. + \mu_B \int_{\mathbb{R}} \left( b_{\alpha}^{\dagger} B_{\alpha}(k) + B_{\alpha}^{\dagger}(k) b_{\alpha} \right) dk \right].$$

All the operators in $H$ obey CAR:

$$\{a_{\alpha}, a_{\beta}^{\dagger}\} = \{b_{\alpha}, b_{\beta}^{\dagger}\} = \delta_{\alpha,\beta}\, \mathbb{1},$$

as well as

$$\left\{ A_{\alpha}(k), A_{\beta}^{\dagger}(q) \right\} = \left\{ B_{\alpha}(k), B_{\beta}^{\dagger}(q) \right\} = \delta_{\alpha,\beta}\, \delta(k-q)\, \mathbb{1}.$$

The other anticommutators are zero.[3] The interpretation of the various contributions in $H$ is very easy. The only new terms are the free Hamiltonians for the reservoirs,

$$\sum_{\alpha} \int_{\mathbb{R}} \Omega_{\alpha}^{A}(k)\, A_{\alpha}^{\dagger}(k) A_{\alpha}(k)\, dk, \qquad \sum_{\alpha} \int_{\mathbb{R}} \Omega_{\alpha}^{B}(k)\, B_{\alpha}^{\dagger}(k) B_{\alpha}(k)\, dk,$$

and the two interaction contributions between the species and their reservoirs,

$$\sum_{\alpha} \mu_A \int_{\mathbb{R}} \left( a_{\alpha}^{\dagger} A_{\alpha}(k) + A_{\alpha}^{\dagger}(k) a_{\alpha} \right) dk,$$
$$\sum_{\alpha} \mu_B \int_{\mathbb{R}} \left( ab_{\alpha}^{\dagger} B_{\alpha}(k) + B_{\alpha}^{\dagger}(k) b_{\alpha} \right) dk.$$

All the constants and the functions appearing here are real-valued, to ensure that $H = H^{\dagger}$.

The Heisenberg equations of motion for the annihilation operators appearing here are the following:

---

[3]A fermionic reservoir is used sometimes in quantum statistical mechanics to avoid problems with unbounded operators (Buffet and Martin, 1978; Martin, 1979). This is a reasonable, technically useful, choice.

$$
\begin{cases}
\dot{a}_\alpha = -i\omega_\alpha^a a_\alpha - i\lambda_\alpha b_\alpha + 2i\gamma_a \sum_\beta p_{\alpha,\beta} a_\beta - i\mu_A \int_{\mathbb{R}} A_\alpha(k), \\
\dot{b}_\alpha = -i\omega_\alpha^b b_\alpha - i\lambda_\alpha a_\alpha + 2i\gamma_b \sum_\beta p_{\alpha,\beta} b_\beta - i\mu_B \int_{\mathbb{R}} B_\alpha(k), \\
\dot{A}_\alpha(k) = -i\Omega_\alpha^A(k) A_\alpha(k) - i\mu_A a_\alpha, \\
\dot{B}_\alpha(k) = -i\Omega_\alpha^B(k) A_\alpha(k) - i\mu_B b_\alpha,
\end{cases} \tag{4.21}
$$

They clearly extend those in Equation 4.11, which are essentially recovered if $\mu_A = \mu_B = 0$. The same techniques used in Section 3.4, assuming in particular that $\Omega_\alpha^A(k) = \Omega_\alpha^A k$ and $\Omega_\alpha^B(k) = \Omega_\alpha^B k$, $\Omega_\alpha^A$, $\Omega_\alpha^B > 0$, allow one to write the first two equations in (4.21) as follows:

$$
\begin{cases}
\dot{a}_\alpha = -\nu_\alpha^a a_\alpha - i\lambda_\alpha b_\alpha + 2i\gamma_a \sum_\beta p_{\alpha,\beta} a_\beta - i\mu_A f_\alpha^A(t), \\
\dot{b}_\alpha = -\nu_\alpha^b b_\alpha - i\lambda_\alpha a_\alpha + 2i\gamma_b \sum_\beta p_{\alpha,\beta} b_\beta - i\mu_B f_\alpha^B(t),
\end{cases} \tag{4.22}
$$

where we have introduced $\nu_\alpha^a := i\omega_\alpha^a + \frac{\pi\mu_A^2}{\Omega_\alpha^A}$, $\nu_\alpha^b := i\omega_\alpha^b + \frac{\pi\mu_B^2}{\Omega_\alpha^B}$, and

$$
f_\alpha^A(t) = \int_{\mathbb{R}} A_\alpha(k)\, e^{-i\Omega_\alpha^A k t}\, dk, \qquad f_\alpha^B(t) = \int_{\mathbb{R}} B_\alpha(k)\, e^{-i\Omega_\alpha^B k t}\, dk.
$$

Here, we consider only the situation in which $\nu_\alpha^a = \nu_\alpha^b =: \nu$ for all $\alpha$. In this way, with the change of variable $x_\alpha(t) := a_\alpha(t)e^{\nu t}$, $y_\alpha(t) := b_\alpha(t)e^{\nu t}$, after some computations we deduce an extended version of Equation 4.15,

$$
\dot{X}_{L^2} = iM_{L^2} X_{L^2} - iF_{L^2}(t), \qquad M_{L^2} = 2\tilde{\gamma} M_{L^2} - \lambda J_{L^2}, \tag{4.23}
$$

where the transpose of $F_{L^2}(t)$ is $(\mu_A f_1^A(t), \ldots, \mu_A f_{L^2}^A(t), \mu_B f_1^B(t), \ldots, \mu_B f_{L^2}^B(t))\, e^{\nu t}$. The differential equation in (4.23) can be solved explicitly, and the solution is

$$
X_{L^2}(t) = e^{iM_{L^2}t}\left( c - i\int dt\, F_{L^2}(t)e^{-iM_{L^2}t} \right), \tag{4.24}
$$

where $c$ has to be fixed by the initial condition $X_{L^2}(0) = X_{L^2}$. The possible existence of an overall decay in the time evolution of, say, the number operator $\hat{n}_\alpha(t) = a_\alpha^\dagger(t)a_\alpha(t)$, can be deduced by writing

$$
a_\alpha^\dagger(t)a_\alpha(t) = e^{-(\nu_\alpha + \overline{\nu_\alpha})t}\, x_\alpha^\dagger(t)x_\alpha(t) = e^{-\frac{2\pi\gamma^2}{\Omega}t}\, x_\alpha^\dagger(t)x_\alpha(t),
$$

so that, if we consider the number operator for the related population, $N^{(a)}(t) := \sum_{\alpha} a_{\alpha}^{\dagger}(t)a_{\alpha}(t)$, this can be written as

$$N^{(a)}(t) := e^{-\frac{2\pi\gamma^2}{\Omega}t}\left(\sum_{\alpha} x_{\alpha}^{\dagger}(t)x_{\alpha}(t)\right).$$

We claim that the exponential $e^{-\frac{2\pi\gamma^2}{\Omega}t}$ is what is responsible for the decay of the populations during the time evolution. In other words, we are saying that the operators $x_{\alpha}^{\dagger}(t)x_{\alpha}(t)$ do not *explode* (in norm) when $t$ increases. For this reason, $\tilde{R}$ can be seen as the *mathematical realization* of the various problems that a migrating population may encounter on its way: bad weather conditions, lack of food, various diseases, and so on.

Of course, a more detailed and satisfactory conclusion concerning the effect of the reservoir could be carried out using Equation 4.24 and repeating essentially the same steps as in Section 3.4. This detailed analysis is beyond our present interests and will not be considered here. However, more detailed computations on the time evolution of a *biological open system* are given in the next chapter, where a model for bacteria is proposed.

## 4.5  COMPETITION BETWEEN POPULATIONS

The same Hamiltonian $H$ introduced in Section 4.3 can be used in the description of competition between two populations, $S_1$ and $S_2$. The equations for the two populations are again those in Equation 4.11. The values of the $p_{\alpha,\beta}$ are chosen exactly as in the previous migration model: an element of $S_1$ or $S_2$ can only move between neighboring cells. Hence, the equation of motion can be written as in Equation 4.18, and the solution is given by Equation 4.19. The difference consists here in the choice of the parameters and of the initial conditions. In fact, in this case, we are no longer interested in having, at $t = 0$, the two populations localized in different regions of $\mathcal{R}$. On the contrary, we consider here again a square lattice, with $L = 11$, but in which both $S_1$ and $S_2$ are localized in a central region $\mathcal{R}_c$ of three-by-three cells, so that they are forced to interact between them from the very beginning, and, also because of this interaction, they start moving along the lattice.

In Figure 4.10, in each row, we plot the local densities of $S_1$ (solid line) and $S_2$ (dashed line) inside (a,c,e) and outside (b,d,f) $\mathcal{R}_c$. Different rows correspond to different values of $\omega_{\alpha}^b$, although all the other parameters coincide. We have chosen two significantly different values of $\gamma_a$ and $\gamma_b$ to give the two populations different mobilities: as $\gamma_a = 0.1 \gg \gamma_b = 0.004$,

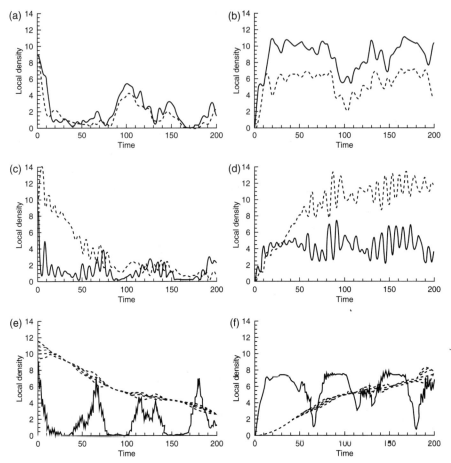

**Figure 4.10**    Evolution of local densities (solid line for $S_1$ and dashed line for $S_2$). Inside $\mathcal{R}_c$ (a,c,e), Outside $\mathcal{R}_c$ (b,d,f). $\gamma_a = 0.1$, $\gamma_b = 0.004$, $\lambda_\alpha = 0.2$ for $\alpha \in \mathcal{R}$. (a,b) $\omega_\alpha^a = 1$, $\omega_\alpha^b = 0.3$, $\forall \alpha \in \mathcal{R}$. (c,d) $\omega_\alpha^a = 1$, $\omega_\alpha^b = 1$, $\forall \alpha \in \mathcal{R}$. (e,f) $\omega_\alpha^a = 1$, $\omega_\alpha^b = 3$, $\forall \alpha \in \mathcal{R}$.

$S_1$ is expected to move faster than $S_2$, and this is exactly what we see in the plots.

Moreover, in analogy with what was discussed before, we expect that $\omega_\alpha^a$ and $\omega_\alpha^b$ play the role of inertia of the two populations in the different cells, so that we expect that changing the ratio $\frac{\omega_\alpha^a}{\omega_\alpha^b}$, or the difference $\left| \omega_\alpha^a - \omega_\alpha^b \right|$, will produce a change in the relative behavior of $S_1$ and $S_2$. Both these features are clearly displayed in Figure 4.10: $S_1$ tends to move away from $\mathcal{R}_c$ faster (or even much faster) than $S_2$. Moreover, going from the first row ($\omega_\alpha^a = 1$, $\omega_\alpha^b = 0.3$, $\forall \alpha \in \mathcal{R}$) to the last one ($\omega_\alpha^a = 1$, $\omega_\alpha^b = 3$, $\forall \alpha \in \mathcal{R}$), it is clear that $S_2$ moves away from $\mathcal{R}_c$ slower and slower: its inertia increases together with the value of $\omega_\alpha^b$.

Particularly interesting is the second row, where the density of $S_2$ in $R_c$ first increases very fast, while that of $S_1$ decreases: this can be considered as the evidence of a bigger *efficiency* of $S_2$ compared with that of $S_1$, which is forced by $S_2$ to leave $R_c$. For instance, thinking of $S_1$ as *preys* and of $S_2$ as *predators*, we can say that the preys run very fast away from the region where the predators stay localized. Hence, $\gamma_a$ and $\gamma_b$ can be considered here, other than as diffusion coefficients, as measuring a sort of *inverse ability* of the two populations: as $\gamma_b^{-1} \gg \gamma_a^{-1}$, $S_2$ is much stronger than $S_1$, and the preys are killed significantly by the predators or, if they survive, run away from $R_c$ very quickly.

Finally, due to the absence of a reservoir, no decay is expected for large $t$, and this is exactly what the plots show. In order to have such a decay, the reservoir must be considered inside the model, as we have already seen that such a reservoir could mimic all the *troubles* that $S_1$ and $S_2$ may experience other than the mutual interaction (lack of food, other predators, cold winters, hot summers, etc.).

## 4.6   FURTHER COMMENTS

At http://www.unipa.it/fabio.bagarello/ricerca/videos.htm, the interested reader can find few simple animations concerning migration, Movies 1 and 2, and competition between species, Movies 3 and 4. These movies show, in two windows, the behavior of the densities of the species $S_1$ (left) and $S_2$ (right) in the lattice $R$. The higher the density of the species, the lighter the related region. In all the movies we have taken $\gamma_a = 0.1$ and $\gamma_b = 0.004$, so that $\gamma_a > \gamma_b$. This has a first immediate consequence: in all the plots, we see that $S_1$ moves much faster than $S_2$, whose reaction appears rather slow when compared to $S_1$. Moreover, at least in the behavior of the *fast* species, in the time interval considered we can see a quasiperiodic behavior: $S_1$ goes back and forth from the region where it is originally (i.e., at $t = 0$) localized, even if we never recover exactly the starting configuration. More in detail, in Movie 1, where $S_1$ and $S_2$ are localized at $t = 0$ respectively in $R_1$ and $R_2$, it is clear that $S_1$ moves toward $R_2$, and then goes back (the welfare is reached!) while different members of $S_1$ try again to reach $R_2$. Note that there exist intervals of time during which $S_1$ is partly localized in $R_1$, partly in $R_2$, and partly on the way, in $R_3$: someone is back in Africa, some other is (hopefully) integrated in Europe, and still some other is trying to reach Europe. As for $S_2$, both Movie 1 and Movie 2 show that the inertia of this second population is very high. This is in agreement with the values chosen for the omegas and with their meaning: while in Movie 1

$\omega_\alpha^a = \omega_\alpha^b = 1$ for all possible $\alpha$ (so that the difference in the mobility of $S_1$ and $S_2$ is only due to $\gamma_a$ and $\gamma_b$), in Movie 2 $\omega_\alpha^a = 1$, while $\omega_\alpha^b = 2$ for all $\alpha$: this causes the inertia of $S_2$ to increase, in agreement with our interpretation of the parameters of the free Hamiltonian.

Movies 3 and 4 show how competition between species works in our framework. In this case, the two species are, at $t = 0$, localized in the central region $\mathcal{R}_c$ of our lattice. Then, because of the interaction between the two, they move away from $\mathcal{R}_c$. In particular, in Movie 3 we see that $S_1$ moves faster (but not so much) than $S_2$. In Movie 4, there is a big difference in speed: it takes much longer for $S_2$ to (partially) leave $\mathcal{R}_c$. Again this is due to the fact that, while in Movie 3, we have taken $\omega_\alpha^a = 1$ and $\omega_\alpha^b = 0.3$, for all $\alpha$, in Movie 4 we have taken $\omega_\alpha^a = 1$ and $\omega_\alpha^a = 3$. In particular, Movie 3 shows that what is particularly relevant to determine the mobility of a certain species is, first of all, the value of the related diffusion coefficients $\gamma$s in $H_{\text{diff}}$, while the values of the omegas become relevant only at the second level.

We end this chapter with some bibliographic notes. The mathematically oriented literature on migration is very poor. To our knowledge, only a few papers deal with this problem. For instance, in Quint and Shubik (1994), the authors propose a game-theoretic model of migration of animals, while in Bijwaard (2008), the author concentrates his attention on The Netherlands, distinguishing the role of migrants going back to their own countries and those remaining in The Netherlands. These different behaviors, as we have seen, are also recovered in our model. We want to add that, in our knowledge, the one proposed here is the first somehow quantitative model describing migration. Of course, our model could be improved in many ways. In particular, we could try to implement some decision-making mechanism, as those discussed in Yukalov and Sornette (2009a,b, 2012): why and when does someone decide to leave his own country? How does such a decision take a concrete form? These are quite fascinating questions, for which we have, at the time being, no answer!

Exactly on the opposite side, the literature on competition between species is huge. In Hanski (1983, 2008), for instance, the author is interested mainly in the possibility of deducing some equilibrium between different species in a patchy environment, and the author finds conditions for such an equilibrium to be reached. In Hanski (1983), moreover, the author proposes what he calls a *predictor of success in local competition*. Something similar is considered also in this chapter as we have seen that the pair $(\gamma, \omega_\alpha)$ gives a very good indication of what the dynamics will be and which will be the *winning* species. In Comins and Hassel (1987), the dynamics of two competing species spatially distributed is considered, adding to these species a shared natural enemy moving in the patchy environment. In particular, the authors have considered two different kinds of enemies: a generalist predator, whose dynamics is uncoupled from those of the two prey species, and a specialist (e.g., a parasitoid), whose dynamics is entirely coupled to those of its two preys. This looks like an extended version of our simpler system, which could also be considered adopting our framework simply by adding a third fermionic operator in our model. Another possible suggestion, coming from Comins and Hassel (1987) and Ives and May (1985), concerns the role of the initial conditions and of the diffusion coefficients $p_{\alpha,\beta}$: these papers suggest that the distributions of the species at $t = 0$ may have some important effect on their dynamics. Moreover, the difference between the cells of the lattice may also produce something interesting. Then the future is signed: we are interested in changing the initial conditions, considering species originally localized in $\mathcal{R}$ in different, physically relevant ways, and we are further interested in taking $p_{\alpha,\beta}$, with $\alpha$ and $\beta$ nearest neighbors, somehow randomly, to model the possibility that some cell of $\mathcal{R}$ is *less favorable* (less food, colder, hotter, etc.) than others.

# CHAPTER 5

# LEVELS OF WELFARE: THE ROLE OF RESERVOIRS

In recent years, a new branch of biology has started to attract the interest of several scientists, the so-called quantum biology. The two main experimental/theoretical facts that suggest the relevance of quantum mechanics in macroscopic biological systems are photosynthesis (Engel et al., 2007; Panitchayangkoon et al., 2011) and the ability of some birds to navigate somehow guided by the Earth's magnetic field (Gauger et al., 2011). A recent review paper is due to Ball (2011), where these and other applications are briefly described. This chapter might be considered as a further, may be more quantitative, example of how quantum mechanics, or quantum mechanical tools, could be used in biology. Here, we consider a biological-like model of *welfare*, in which a system $S$ of entities (bacteria, animals, or human beings, for instance) is described by different levels of welfare (LoWs) describing their overall status as the higher the level, the better their living conditions. On the other hand, those elements of $S$, which occupy the lowest LoW, are very close to dying! In a certain sense, the general strategy is not very different from the one adopted for love stories: discrete levels are used to describe different statuses of the entities that compose the system.

In the previous chapters, we have considered the role of the reservoir: how this affects $S$, whereas in this chapter, we are also interested in the

*Quantum Dynamics for Classical Systems: With Applications of the Number Operator*,
First Edition. Fabio Bagarello.
© 2013 John Wiley & Sons, Inc. Published 2013 by John Wiley & Sons, Inc.

dynamical behavior of the reservoir itself because, as we discuss in a moment, it will acquire a sharp biological interpretation.

## 5.1   THE MODEL

Our main aim is to describe a nonconservative system, in which the total number of *entities* is not necessarily preserved during the time evolution. In particular, we want to consider the possibility that some entity of $S$ dies, and that some other is born while the system evolves. In view of what has been discussed in Section 2.7.1, it does not seem to be a good idea to look for a non-self-adjoint Hamiltonian to describe such a system: the standard Heisenberg dynamics is lost, and it is not completely clear which kind of differential equation should replace Equation 2.2.[1] The natural choice consists, therefore, in adding some reservoirs interacting with $S$. We see in a moment that, for the system we have in mind, a reasonable choice is the introduction of two reservoirs, having two different meanings.

Let $S$ consist, at $t = 0$, of a certain number of members, which we have called the *entities*, and which can be described by $M + 1$ different LoWs; the ones in the highest level, $M$, live in a wonderful situation, that is, they have food, space, good temperature around, ..., whereas the ones in level $M - 1$ have a slightly worse situation, which is even worse for the entities in level $M - 2$, and so on. Going down and down, we arrive at level 0, where there is almost no food, no space, the temperature is not appropriate, .... We assume that, at $t = 0$, $n_0$ entities are in the zeroth level, $n_1$ in the first level, and so on, up to $n_M$ entities, which are in the $M$th level. Therefore, in the $M + 1$ LoWs of $S$, $N_0 = \sum_{k=0}^{M} n_k$ entities are distributed at $t = 0$. During the *life* of each entity of $S$, some garbage is produced, filling up what we call $\mathcal{R}_g$, the *reservoir of the garbage*. Moreover, if some entity dies (and this can occur in our model only if that entity was occupying the zeroth level), the garbage increases as well.

Needless to say, $\mathcal{R}_g$ is not sufficient here. We need to add the *reservoir of the food*, $\mathcal{R}_f$, to the game, in order for the entities to grow up. Of course, it is not difficult to extend the meaning of $\mathcal{R}_f$ and $\mathcal{R}_g$ so as to attach to them other possible effects: for instance, $\mathcal{R}_g$ could be used not only to represent the garbage, but also to represent the *enemies* of the entities of

---

[1]Recall that, anyhow, a reasonable possibility was proposed in Section 2.7.1.

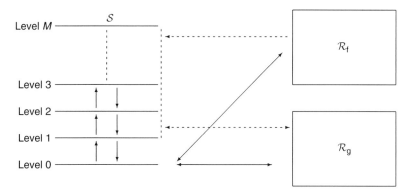

**Figure 5.1**    The system coupled to two reservoirs.

$S$. However, in this chapter, we focus on this simpler and rather natural interpretation.

Summarizing, $\mathcal{R}_f$ provides food to $S$, improving the *quality of life* of its entities. On the other hand, $\mathcal{R}_g$ produces a disturbing effect that worsens the quality of life of the members of $S$. The situation is schematically described in Figure 5.1.

In particular, this figure describes the following facts:

1. The food is needed by all the entities of $S$, belonging to each level, $0, 1, \ldots, M$. This is the meaning of the arrows going from $\mathcal{R}_f$ to the various levels of $S$.

2. During their lives, the entities of levels $1, 2, \ldots, M$ produce some garbage, filling up $\mathcal{R}_g$. This is why we have arrows going from levels $1, 2, \ldots, M$ to $\mathcal{R}_g$.

3. $\mathcal{R}_g$ can also be filled up by the entities occupying level 0 during their life or, in part, when they die.

4. Part of these dying entities in level 0 becomes food for the other members of $S$, so they also fill up, in part, $\mathcal{R}_f$.

5. The garbage influences the welfare of the entities of $S$, decreasing the quality of their lives. This is why we have also arrows going from $\mathcal{R}_g$ to each level of $S$.

6. $\mathcal{R}_g$ is coupled in a different way with levels $1, 2, \ldots, M$ and with level 0. The reason for this difference is that only the entities in level 0 can die. The entities in level $j$, with $j \geq 1$ and $j < M$, cannot die; they can improve their LoW moving to level $j + 1$, or they can go down to level $j - 1$, depending on their interactions with $\mathcal{R}_f$ and $\mathcal{R}_g$.

We are now ready to write the Hamiltonian for our open system $\tilde{S} = S \cup \mathcal{R}_f \cup \mathcal{R}_g$, following the main ideas already discussed in Chapters 3 and 4. We say much more on these *constructing techniques* in the next chapter. We put

$$\begin{cases} H = H_0 + H_{\text{int}}, \\ H_0 = H_{0,S} + H_{0,\mathcal{R}_f} + H_{0,\mathcal{R}_g}, \\ H_{\text{int}} = H_{\text{int,f}} + H_{\text{int,g}} + H_{\text{jump}}, \end{cases} \quad (5.1)$$

where $H_0$ is the free Hamiltonian of $\tilde{S}$, whereas $H_{\text{int}}$ contains all the interactions between its various ingredients. We adopt bosonic operators in description of both the system, $S$, and the reservoirs, $\mathcal{R}_f$ and $\mathcal{R}_g$. With obvious notation, we assume the following commutation relations:

$$[a_k, a_l^{\dagger}] = \delta_{k,l} \mathbb{1}, \quad [B_f(k), B_f^{\dagger}(q)] = \delta(k - q) \mathbb{1},$$

$$[B_{g(k)}, B_g^{\dagger}(q)] = \delta(k - q) \mathbb{1}, \quad (5.2)$$

while all the other commutators are zero. Various ingredients of the free Hamiltonian are the following:

$$\begin{cases} H_{0,S} = \sum_{k=0}^{M} \omega_k a_k^{\dagger} a_k, \\ H_{0,\mathcal{R}_f} = \int_{\mathbb{R}} \Omega_f(k) B_f^{\dagger}(k) B_f(k) \, dk, \\ H_{0,\mathcal{R}_g} = \int_{\mathbb{R}} \Omega_g(k) B_g^{\dagger}(k) B_g(k) \, dk, \end{cases} \quad (5.3)$$

which are such that, when no interaction exists, the number operators of the system and of the reservoirs,

$$\hat{N}_S := \sum_{k=0}^{M} a_k^{\dagger} a_k, \quad \hat{N}_f := \int_{\mathbb{R}} B_f^{\dagger}(k) B_f(k) \, dk,$$

$$\hat{N}_g := \int_{\mathbb{R}} B_g^{\dagger}(k) B_g(k) \, dk, \quad (5.4)$$

stay all constant in time. This is in agreement with **Rule 2** of Chapter 3.[2] Notice that the $\omega_l$s are real numbers and that $\Omega_f(k)$ and $\Omega_g(k)$ are real valued functions of $k$, in order $H_0$ to be self-adjoint. Of course,

---

[2]See also **R1** in Chapter 6.

this means that the $n_0$ entities in level 0 are described by the vector $\varphi_{n_0}^{(0)} := \frac{1}{\sqrt{n_0!}} (a_0^\dagger)^{n_0} \varphi_0^{(0)}$, the $n_1$ entities in level 1 are described by the vector $\varphi_{n_1}^{(1)} := \frac{1}{\sqrt{n_1!}} (a_1^\dagger)^{n_1} \varphi_0^{(1)}$, and so on. Here $\varphi_0^{(k)}$ is the *vacuum* of the operator $a_k$: $a_k \varphi_0^{(k)} = 0$, for $k = 0, 1, \ldots, M$. The vector describing $\mathcal{S}$ at $t = 0$ is the following tensor product:

$$\varphi_{n_0,n_1,\ldots,n_M} = \varphi_{n_0}^{(0)} \otimes \varphi_{n_1}^{(1)} \otimes \cdots \otimes \varphi_{n_M}^{(M)}$$

$$= \frac{1}{\sqrt{n_0! \, n_1! \ldots n_M!}} (a_0^\dagger)^{n_0} (a_1^\dagger)^{n_1} \cdots (a_M^\dagger)^{n_M} \varphi_0, \qquad (5.5)$$

where $\varphi_0 = \varphi_0^{(0)} \otimes \varphi_0^{(1)} \otimes \cdots \otimes \varphi_0^{(M)}$ is the vacuum of the system.

Regarding $H_{\text{int}}$ in Equation 5.1, let us first consider $H_{\text{int,f}}$ and $H_{\text{int,g}}$. They are defined as follows:

$$\begin{cases} H_{\text{int,f}} = \lambda_f \int_{\mathbb{R}} \left( a_0 B_f^\dagger(k) + a_0^\dagger B_f(k) \right) dk, \\ H_{\text{int,g}} = \lambda_g \int_{\mathbb{R}} \left( a_0 B_g^\dagger(k) + a_0^\dagger B_g(k) \right) dk, \end{cases} \qquad (5.6)$$

where $\lambda_f$ and $\lambda_g$ are real constants, measuring the interaction strength between the two reservoirs and the entities in level 0. The meaning of $H_{\text{int},f}$ is the following: if the number of entities in level 0 increases, food decreases because this is used by these new entities. This effect is described by the term $a_0^\dagger B_f(k)$. The adjoint contribution, $a_0 B_f^\dagger(k)$, describes the fact that part of the entities of $\mathcal{S}$, which dies (of course we are referring here to those entities which were in level 0 before dying), becomes food for the other entities and, as a consequence, they contribute to $\mathcal{R}_f$. A similar contribution, $a_0 B_g^\dagger(k)$, appears in $H_{\text{int},g}$. This term models the fact that dying produces garbage as well, and not only food! The other contribution in $H_{\text{int},g}$, $a_0^\dagger B_g(k)$, is mainly added to ensure self-adjointness of the Hamiltonian, but its effect is *erased* by adopting convenient initial conditions, that is, requiring that, at $t = 0$, $\mathcal{R}_g$ is empty. This implies that, at least for small values of $t$, $\mathcal{R}_g$ can only *receive* and not *give* anything: the double solid arrow between level 0 and $\mathcal{R}_g$ in Figure 5.1 produces, because of these initial conditions, a unidirectional flow from level 0 to $\mathcal{R}_g$.

The last contribution to $H_{\text{int}}$ in Equation 5.1 is the Hamiltonian $H_j$, where j stands for *jump*: $H_j$ is responsible for the possible changes of

LoW of the entities belonging to the system $S$. We begin with introducing the *jump up* operator

$$X_+ := \sum_{k=0}^{M-1} a_k a_{k+1}^\dagger. \tag{5.7}$$

The reason for this name is the following: if we act with $X_+$ on, for example, the vector $\varphi_{0,1,0,0,...,0}$, we produce a new vector, which is proportional to $\varphi_{0,0,1,0,...,0}$. Therefore, the entity, originally in level 1, has moved to level 2. For obvious reasons, its adjoint $X_- := (X_+)^\dagger = \sum_{k=0}^{M-1} a_k^\dagger a_{k+1}$, is called the *jump down* operator. In Figure 5.1, these operators are described by the up and down arrows between the different levels of $S$. Now we define

$$H_j = \lambda \int_{\mathbb{R}} \left( X_+ B_f(k) B_g^\dagger(k) + X_- B_f^\dagger(k) B_g(k) \right) dk. \tag{5.8}$$

The first contribution, $X_+ B_f(k) B_g^\dagger(k)$, describes the fact that some food has been used to increase the number of entities in higher LoWs and that, at the same time, more garbage is produced. This can be easily understood, for instance, in a human context: rich populations produce higher amount of garbage than the poor ones, and they also consume more food. The second contribution in $H_j$, $X_- B_f^\dagger(k) B_g(k)$ is needed essentially to ensure self-adjointness of $H$. Strictly speaking, this corresponds to the following phenomenon: the LoWs of the entities of $S$ decrease, whereas the food of $\mathcal{R}_f$ increases and the garbage decreases as well. Then, its biological meaning is the following: when the quality of life of a given entity decreases, it needs a smaller amount of food to live (or to survive). But, during the time evolution, as $\mathcal{R}_f$ continues being filled up by the other dying entities from level 0, it is as if some net extra food appears in $\mathcal{R}_f$. Moreover, as that entity produces less garbage than before, it is as if $\mathcal{R}_g$ is being emptied out. In different words, this term in $H_j$ is better understood when we consider the deviations from the average values of the quantities involved in the description of the reservoirs.

It is not surprising that $[H, \hat{N}_S] \neq 0$: entities can die or be born, in $S$, so that their total number is not expected to be an integral of motion. On the other hand, if we define the *global number operator*

$$\hat{N} = \hat{N}_S + \hat{N}_f + \hat{N}_g, \tag{5.9}$$

we can check that

$$[H, \hat{N}] = 0. \tag{5.10}$$

This implies that $\frac{d}{dt}\hat{N}(t) = 0$, and therefore, that

$$\frac{d}{dt}\hat{N}_S(t) = -\frac{d}{dt}\left(\hat{N}_f(t) + \hat{N}_g(t)\right).$$

Hence, the time variation of the entities of $\mathcal{S}$ is balanced by the variations of the two reservoirs, considered together. The relevant equations of motion for $\tilde{\mathcal{S}}$ are obtained using Equation 2.2:

$$\begin{cases} \dot{a}_0(t) = -i\omega_0 a_0(t) - i\lambda_f \int_{\mathbb{R}} B_f(k,t)\,dk - i\lambda_g \int_{\mathbb{R}} B_g(k,t)\,dk + \\ \qquad -i\lambda a_1(t) \int_{\mathbb{R}} B_f^\dagger(k,t) B_g(k,t)\,dk, \\ \dot{a}_j(t) = -i\omega_j a_j(t) - i\lambda \int_{\mathbb{R}} \left(a_{j-1}(t) B_f(k,t) B_g^\dagger(k,t)\right. \\ \qquad \left. + a_{j+1}(t) B_f^\dagger(k,t) B_g(k,t)\right)\,dk, \\ \dot{a}_M(t) = -i\omega_M a_M(t) - i\lambda a_{M-1}(t) \int_{\mathbb{R}} B_f(k,t) B_g^\dagger(k,t)\,dk, \\ \dot{B}_f(k,t) = -i\Omega_f(k) B_f(k,t) - i\lambda_f a_0(t) - i\lambda X_-(t) B_g(k,t), \\ \dot{B}_g(k,t) = -i\Omega_g(k) B_g(k,t) - i\lambda_g a_0(t) - i\lambda X_+(t) B_f(k,t), \\ \dot{X}_+(t) = i\varepsilon X_+(t) - i\lambda_f a_1^\dagger(t) \int_{\mathbb{R}} B_f(k,t)\,dk \\ \qquad -i\lambda_g a_1^\dagger(t) \int_{\mathbb{R}} B_g(k,t)\,dk + +i\lambda \left(a_0^\dagger(t) a_0(t) - a_M^\dagger(t) a_M(t)\right) \\ \qquad \int_{\mathbb{R}} B_f^\dagger(k,t) B_g(k,t)\,dk, \end{cases}$$

$$(5.11)$$

where $j = 1, 2, \ldots, M-1$ and $\varepsilon := \omega_k - \omega_{k-1}$, $k = 1, 2, \ldots, M$. In particular, we are assuming here, to simplify the treatment a little bit, that the *energy* levels of $\mathcal{S}$ are all equally spaced. Together with their adjoints, this set of equations would produce a closed set of nonlinear differential equations, whose exact solution, however, is very hard to find. Nevertheless, interesting results can be deduced if we assume that the *strongest* interaction in the game is the one between level 0 and the two reservoirs, while the other levels interact with $\mathcal{R}_f$ and $\mathcal{R}_g$ *weakly*. As one could imagine from the above definition of the Hamiltonian of $\tilde{\mathcal{S}}$, see Equations 5.1, 5.6, and 5.8, this is reflected by the following requirement:

$$\lambda \ll \min\left(\lambda_f, \lambda_g\right). \qquad (5.12)$$

## 5.2 THE SMALL λ REGIME

Before considering this approximation, the following remark is in order: the equations in (5.11) clearly show the difference between the *extreme* and *intermediate* states of $S$. This difference is motivated by the fact that if $j = 1, 2, \ldots, M - 1$, the entities may go up and down to the $j + 1$ and $j - 1$ LoW, respectively; if $j = 0$, they can only go to level 1 (or die!), and if $j = M$, they can only go to level $M - 1$. This aspect is also maintained in the approximated version of the equation (5.11), deduced by adopting Equation 5.12, and, which are the following:

$$
\begin{cases}
\dot{a}_0(t) = -i\omega_0 a_0(t) - i\lambda_{\mathrm{f}} \int_{\mathbb{R}} B_{\mathrm{f}}(k, t)\, dk - i\lambda_{\mathrm{g}} \int_{\mathbb{R}} B_{\mathrm{g}}(k, t)\, dk \\[2mm]
\dot{a}_j(t) = -i\omega_j a_j(t) - i\lambda \int_{\mathbb{R}} \Big( a_{j-1}(t) B_{\mathrm{f}}(k, t) B_{\mathrm{g}}^{\dagger}(k, t) \\[2mm]
\qquad\qquad + a_{j+1}(t) B_{\mathrm{f}}^{\dagger}(k, t) B_{\mathrm{g}}(k, t) \Big)\, dk, \\[2mm]
\dot{a}_M(t) = -i\omega_M a_M(t) - i\lambda a_{M-1}(t) \int_{\mathbb{R}} B_{\mathrm{f}}(k, t) B_{\mathrm{g}}^{\dagger}(k, t)\, dk, \\[2mm]
\dot{B}_{\mathrm{f}}(k, t) = -i\Omega_{\mathrm{f}}(k) B_{\mathrm{f}}(k, t) - i\lambda_{\mathrm{f}} a_0(t), \\[2mm]
\dot{B}_{\mathrm{g}}(k, t) = -i\Omega_{\mathrm{g}}(k) B_{\mathrm{g}}(k, t) - i\lambda_{\mathrm{g}} a_0(t), \\[2mm]
\dot{X}_+(t) = i\varepsilon X_+(t) - i\lambda_{\mathrm{f}} a_1^{\dagger}(t) \int_{\mathbb{R}} B_{\mathrm{f}}(k, t)\, dk \\[2mm]
\qquad\qquad - i\lambda_{\mathrm{g}} a_1^{\dagger}(t) \int_{\mathbb{R}} B_{\mathrm{g}}(k, t)\, dk.
\end{cases}
\tag{5.13}
$$

It is interesting to stress that our approximation does not eliminate completely the role of $H_j$, as the terms proportional to $\lambda$ still appear in the equations for $a_j(t)$ and $a_M(t)$. This is a good indication that we are not going to make our model trivial. In other words, Equation 5.13 is much better than simply taking $\lambda = 0$ in the definition of $H$, of course. Moreover, and very important, the last equation in Equation 5.13 does not play any role in the solution of the other ones. In fact, if we consider the first, fourth, and fifth equations, they already form a closed system of differential equations by themselves, so that they can be solved (in principle), giving $a_0(t)$, $B_{\mathrm{f}}(k, t)$, and $B_{\mathrm{g}}(k, t)$. These solutions can now be introduced in the second and third equations given earlier, getting $a_j(t)$ for $j = 1, 2, \ldots, M$, and the related number operators in the usual way: $\hat{n}_j(t) = a_j^{\dagger}(t) a_j(t)$, $j = 0, 1, \ldots, M$. The average values of these operators, see later, will finally produce the time evolution of the populations of the entities of $S$ in the various LoWs.

### 5.2.1  The Sub-Closed System

We focus now on the following subsystem of Equation 5.13:

$$
\begin{cases}
\dot{a}_0(t) = -i\omega_0 a_0(t) - i\lambda_f \int_{\mathbb{R}} B_f(k, t)\, dk - i\lambda_g \int_{\mathbb{R}} B_g(k, t)\, dk \\
\dot{B}_f(k, t) = -i\Omega_f(k) B_f(k, t) - i\lambda_f a_0(t), \\
\dot{B}_g(k, t) = -i\Omega_g(k) B_g(k, t) - i\lambda_g a_0(t),
\end{cases}
\tag{5.14}
$$

which is already closed. In particular, the (formal) solutions of the last two equations, satisfying the initial conditions $B_f(k, 0) = B_f(k)$ and $B_g(k, 0) = B_g(k)$, are the following:

$$
\begin{cases}
B_f(k, t) = B_f(k)\, e^{-i\Omega_f(k)t} - i\lambda_f \int_0^t a_0(t_1)\, e^{-i\Omega_f(k)(t-t_1)}\, dt_1, \\
B_g(k, t) = B_g(k)\, e^{-i\Omega_g(k)t} - i\lambda_g \int_0^t a_0(t_1)\, e^{-i\Omega_g(k)(t-t_1)}\, dt_1.
\end{cases}
\tag{5.15}
$$

Now, repeating (more or less) the same procedure outlined in Section 2.7, we choose the usual linear expressions for the functions $\Omega_f(k)$ and $\Omega_g(k)$: $\Omega_f(k) = \Omega_f k$, $\Omega_g(k) = \Omega_g k$, with $\Omega_f$ and $\Omega_g$ real and positive constants. With these choices, and with the usual tricks, the operators in Equation 5.15 produce the following results:

$$
\begin{cases}
\int_{\mathbb{R}} B_f(k, t)\, dk = b_f(t) - i\lambda_f \dfrac{\pi}{\Omega_f}\, a_0(t), \\
\int_{\mathbb{R}} B_g(k, t)\, dk = b_g(t) - i\lambda_g \dfrac{\pi}{\Omega_g}\, a_0(t),
\end{cases}
\tag{5.16}
$$

where we have defined the operators:

$$
b_f(t) := \int_{\mathbb{R}} B_f(k)\, e^{-i\Omega_f k t}\, dk, \qquad b_g(t) := \int_{\mathbb{R}} B_g(k)\, e^{-i\Omega_g k t}\, dk. \tag{5.17}
$$

Replacing Equation 5.16 in the first equation of Equation 5.14, we get

$$
\dot{a}_0(t) = -(i\omega_0 + \pi v^2) a_0(t) - i b(t), \tag{5.18}
$$

where we have introduced the following simplifying notation:

$$
b(t) := \lambda_f b_f(t) + \lambda_g b_g(t), \qquad v^2 := \frac{\lambda_f^2}{\Omega_f} + \frac{\lambda_g^2}{\Omega_g}.
$$

The solution of Equation 5.18 can be easily found:

$$a_0(t) = a_0 e^{-(i\omega_0 + \pi v^2)t} - i \int_0^t b(t_1) e^{-(i\omega_0 + \pi v^2)(t - t_1)} \, dt_1, \qquad (5.19)$$

which satisfies the correct initial condition $a_0(0) = a_0$. The adjoint of Equation 5.19 produces the time evolution of the creation operator $a_0^\dagger(t)$. In order to find the time evolution of the population in level 0, we now need to compute the mean value of $\hat{n}_0(t) = a_0^\dagger(t) a_0(t)$ on a state $\langle \rangle$, which is the tensor product of a number state on $\mathcal{S}$, $\langle \rangle_S$, and two reservoir states $\langle \rangle_f$ and $\langle \rangle_g$, which have the following properties, see Section 2.7:

$$
\begin{aligned}
\langle B_f^\dagger(k) B_f(q) \rangle_f &= n_f(k)\, \delta(k - q), \\
\langle B_g^\dagger(k) B_g(q) \rangle_g &= n_g(k)\, \delta(k - q),
\end{aligned}
\qquad (5.20)
$$

where $n_f(k)$ and $n_g(k)$ are two real functions, which, for simplicity, we assume to be constant in $k$: $n_f(k) = n_f$ and $n_g(k) = n_g$.

As for $\langle \rangle_S$, this is the usual number vector state constructed as follows: for each operator $X$ acting on $\mathcal{S}$, we put

$$\langle X \rangle_S = \langle \varphi_{n_0, n_1, \ldots, n_M}, X \varphi_{n_0, n_1, \ldots, n_M} \rangle,$$

where $(n_0, n_1, \ldots, n_M)$ describes the distribution of the population in the various LoWs at time $t = 0$.

We are now ready to compute $n_0(t) := \langle \hat{n}_0(t) \rangle$. Using Equations 5.19 and 5.20 and the earlier definitions, we finally get

$$n_0(t) = n_0 e^{-2\pi v^2 t} + \frac{1}{v^2} \left( \frac{n_f \lambda_f^2}{\Omega_f} + \frac{n_g \lambda_g^2}{\Omega_g} \right) \left( 1 - e^{-2\pi v^2 t} \right). \qquad (5.21)$$

It is clear that $n_0(0) = n_0$. Moreover, for large $t$, we find that

$$\lim_{t, \infty} n_0(t) = \frac{1}{v^2} \left( \frac{n_f \lambda_f^2}{\Omega_f} + \frac{n_g \lambda_g^2}{\Omega_g} \right) =: n_0^\infty,$$

which, recalling the definition of $v^2$, can be written as

$$n_0^\infty = \frac{n_f \lambda_f^2 \Omega_g + n_g \lambda_g^2 \Omega_f}{\lambda_f^2 \Omega_g + \lambda_g^2 \Omega_f}. \qquad (5.22)$$

This formula shows that, if in particular $n_f = n_g$, then $n_0^\infty = n_f$. In other words, if $\mathcal{R}_f$ and $\mathcal{R}_g$ are equally filled at $t = 0$, then the ground level of $\mathcal{S}$ changes the number of its entities from $n_0$ to $n_f$. Depending on the relation between these two numbers, level 0 is emptied out or filled up. In particular, if $n_f = n_g = 0$, level 0 becomes completely empty after some time. We postpone to Section 5.4 a detailed analysis of this and other results, concerning also the populations of the two reservoirs.

### 5.2.2 And Now, the Reservoirs!

Compared with what we have done in Chapters 3 and 4, the analysis of the open system we consider now is, in a sense, more complete as we are also going to investigate in some detail what happens to $\mathcal{R}_f$ and $\mathcal{R}_g$ during the time evolution. The reason for this is in the physical (and biological) interpretation that we have given to $\mathcal{R}_f$ $\mathcal{R}_g$ here, which makes interesting for us to also consider what is happening to the food and to the garbage of the system.

Before doing this, it might be useful to show why $n_f$ and $n_g$ introduced soon after Equation 5.20 give a measure of how much $\mathcal{R}_f$ and $\mathcal{R}_g$ are *filled up* at $t = 0$. For this, it is convenient to rewrite $\hat{N}_f$ as follows:

$$\hat{N}_f = \int_{\mathbb{R}} dk \int_{\mathbb{R}} dq \, B_f^\dagger(k)\delta(k - q) \, B_f(q).$$

However, it should be understood that this is a formal expression, exactly as its original form in Equation 5.4, because of the distributional nature of the operators $B_f^\dagger(k)$ and $B_f(q)$.[3] Now, taking the mean value of $\hat{N}_f$ on a state $\langle\,\rangle_f$, and using Equation 5.20, with a suitable change of variable we get

$$\langle\hat{N}_f\rangle_f = n_f \int_{\mathbb{R}} \delta(k_1) \, dk_1 \int_{\mathbb{R}} \delta(k_2) \, dk_2 = n_f,$$

as expected. Similarly we could check that $\langle\hat{N}_g\rangle_g = n_g$.

The next step here is to compute the operators $B_f(k, t)$ and $B_g(k, t)$. This can be done replacing $a_0(t)$, as given by Equation 5.19, in Equation 5.15 and performing the integrations. To keep these computations simple, we just consider the first nontrivial contributions of this

---

[3] It is well known that two distributions can be, for instance, summed up although, in general, they cannot be multiplied. This problem is even more severe for operator-valued distributions; see Streiter and Wightman (1964), as it happens in Quantum Field Theory. Consequences of this problem appear in Section 5.3.1.

derivation, keeping $O(\lambda_f)$ but neglecting $O(\lambda_f^2)$ and $O(\lambda_f\lambda_g)$. Just to summarize, we are assuming here not only that $\lambda \ll \min(\lambda_f, \lambda_g)$, see Equation 5.12, but also that both $\lambda_f$ and $\lambda_g$ are sufficiently small. Hence, at this perturbation order,

$$B_f(k, t) = e^{-i\,\Omega_f k\,t}\left\{ B_f(k) - i\lambda_f\, a_0 \frac{e^{(i\,\Omega_f k - i\,\omega_0 - \pi v^2)t} - 1}{i(\Omega_f k - \omega_0) - \pi v^2}\right\}. \qquad (5.23)$$

The mean value of $\hat{N}_f(t)$ can be deduced by performing an easy integral of a certain analytical function having two poles in the complex plane. More in detail we have

$$\langle \hat{N}_f(t)\rangle = \int_{\mathbb{R}} dk \int_{\mathbb{R}} dq\, \langle \hat{B}_f^\dagger(k, t)\, B_f(q, t)\rangle \delta(k - q)$$

$$= n_f + n_0\lambda_f^2 \int_{\mathbb{R}} dk\, |\eta_0(k, t)|^2,$$

where

$$\eta_0(k, t) = e^{-i\Omega_f k\,t} \frac{e^{(i\,\Omega_f k - i\,\omega_0 - \pi v^2)t} - 1}{i(\Omega_f k - \omega_0) - \pi v^2}.$$

The function $|\eta_0(k, t)|^2$ has two poles of the first order in $k_\pm = \frac{1}{\Omega_f}(\omega_0 \pm i\pi v^2)$. It is now possible to conclude, using standard complex integration, that

$$n_f(t) := \langle \hat{N}_f(t)\rangle = n_f + \frac{n_0\,\lambda_f^2}{\Omega_f v^2}\left(1 - e^{-2\pi v^2 t}\right). \qquad (5.24)$$

It is clear that $n_f(0) = n_f$. Moreover, in the limit of large $t$, the behavior of $\mathcal{R}_f$ is easily deduced:

$$\lim_{t,\infty} n_f(t) = n_f + n_0 \frac{\lambda_f^2\Omega_g}{\lambda_f^2\Omega_g + \lambda_g^2\Omega_f} =: n_f^\infty. \qquad (5.25)$$

We refer to Section 5.4 for more considerations on these results.

Most of the steps that have produced $n_f(t)$ can also be repeated in the analysis of the time behavior of the second reservoir, $\mathcal{R}_g$. In particular, we find that, with the same kind of approximations,

$$B_g(k, t) = e^{-i\,\Omega_g k\,t}\left\{ B_g(k) - i\lambda_g\, a_0 \frac{e^{|(i\,\Omega_g k - i\,\omega_0 - \pi v^2)t} - 1}{i(\Omega_g k - \omega_0) - \pi v^2}\right\}, \qquad (5.26)$$

and

$$\langle \hat{N}_{\mathrm{g}}(t) \rangle = \int_{\mathbb{R}} \mathrm{d}k \int_{\mathbb{R}} \mathrm{d}q \, \langle \hat{B}_{\mathrm{g}}^{\dagger}(k, t) \, B_{\mathrm{g}}(q, t) \rangle \delta(k - q)$$

$$= n_{\mathrm{g}} + n_0 \lambda_{\mathrm{g}}^2 \int_{\mathbb{R}} \mathrm{d}k \, |\tilde{\eta}_0(k, t)|^2,$$

where

$$\tilde{\eta}_0(k, t) = \mathrm{e}^{-i\Omega_{\mathrm{g}} k t} \frac{\mathrm{e}^{(i\,\Omega_{\mathrm{g}} k - i\,\omega_0 - \pi\, \nu^2) t} - 1}{i(\Omega_{\mathrm{g}} k - \omega_0) - \pi \nu^2}.$$

Therefore,

$$n_{\mathrm{g}}(t) := \langle \hat{N}_{\mathrm{g}}(t) \rangle = n_{\mathrm{g}} + \frac{n_0 \lambda_{\mathrm{g}}^2}{\Omega_{\mathrm{g}} \nu^2} \left( 1 - \mathrm{e}^{-2\pi \nu^2 t} \right). \tag{5.27}$$

It is clear that $n_{\mathrm{g}}(0) = n_{\mathrm{g}}$. Moreover, for large $t$, we obtain

$$\lim_{t, \infty} n_{\mathrm{g}}(t) = n_{\mathrm{g}} + n_0 \frac{\lambda_{\mathrm{g}}^2 \Omega_{\mathrm{f}}}{\lambda_{\mathrm{g}}^2 \Omega_{\mathrm{f}} + \lambda_{\mathrm{f}}^2 \Omega_{\mathrm{g}}} =: n_{\mathrm{g}}^{\infty}. \tag{5.28}$$

In Section 5.4, we see how to introduce a sort of *efficiency test* for $\tilde{\mathcal{S}}$, a test based on a certain relation between the parameters $\lambda_{\mathrm{f}}$, $\Omega_{\mathrm{g}}$, $\lambda_{\mathrm{g}}$, and $\Omega_{\mathrm{f}}$. Here *efficiency* stands for *capability of $\tilde{\mathcal{S}}$ to produce food out of the dying entities of $\mathcal{S}$*.

## 5.3 BACK TO $\mathcal{S}$

We are now ready to write the closed system of differential equations governing the time evolution of the other levels of $\mathcal{S}$. In particular, defining the quadratic operator

$$\beta(t) := \int_{\mathbb{R}} B_{\mathrm{f}}(k, t) \, B_{\mathrm{g}}^{\dagger}(k, t) \, \mathrm{d}k, \tag{5.29}$$

the differential equations for $a_j(t)$, $j = 1, 2, \ldots, M$, in Equation 5.13 can be easily rewritten in matrix form as

$$\dot{A}(t) = -i \, T(t) \, A(t) - i\lambda \beta(t) \, K(t). \tag{5.30}$$

Here, we have introduced the vectors $A(t)$ and $K(t)$ as follows:

$$A(t) := \begin{pmatrix} a_1(t) \\ a_2(t) \\ \vdots \\ \vdots \\ \vdots \\ a_M(t) \end{pmatrix}, \qquad K(t) := \begin{pmatrix} a_0(t) \\ 0 \\ 0 \\ \vdots \\ \vdots \\ 0 \end{pmatrix},$$

and the following tridiagonal matrix $T(t)$

$$T(t) := \begin{pmatrix} \omega_1 & \lambda\beta^\dagger(t) & 0 & 0 & 0 & \cdots & \cdots \\ \lambda\beta(t) & \omega_2 & \lambda\beta^\dagger(t) & 0 & 0 & \cdots & \cdots \\ 0 & \lambda\beta(t) & \omega_3 & \lambda\beta^\dagger(t) & 0 & \cdots & \cdots \\ 0 & 0 & \lambda\beta(t) & \omega_4 & \lambda\beta^\dagger(t) & \cdots & \cdots \\ \cdots & \cdots & \cdots & \cdots & \cdots & \cdots & \cdots \\ \cdots & \cdots & \cdots & \cdots & \cdots & \cdots & \cdots \\ \cdots & \cdots & 0 & 0 & 0 & \lambda\beta(t) & \omega_M \end{pmatrix}.$$

We notice that both $\beta(t)$ and $K(t)$ are known operator-valued functions of time, so that Equation 5.30 appears to be a nonhomogeneous, linear differential equation with time-depending coefficients.

The (formal) solution of this equation can be found easily if $\left[ T(t), \int T(t)\,dt \right] = 0$. In this case, we find

$$A(t) = e^{-i \int T(t)\,dt} \left( c - i\lambda \int dt\beta(t)K(t)\, e^{i \int T(t)\,dt} \right), \qquad (5.31)$$

where $c$ should be fixed by the initial conditions. In principle, we are therefore in the position of deducing the various $a_j(t)$, and the related $\hat{n}_j(t) = a_j^\dagger(t)a_j(t)$, consequently. However, rather than continuing with the details of this computation, we concentrate now on a much more *flexible* but still not trivial situation, that is, the case in which $M = 2$. It is not trivial, first of all, as we have at least one LoW, level 1, which can give rise to movements up (to level 2) and to movements down (to level 0). Furthermore, it is flexible as the computations can all be performed analytically in a rather reasonable way. It is also necessary to mention that, in the biological literature, $M = 2$ is considered a good choice to describe, for instance, bacteria: when the environmental conditions are favorable, bacteria live and flourish. This is level 2, for us. Changing the temperature and the pH value and adding toxic compounds decrease the LoW: bacteria move down to level 1. Where the conditions are very unfavorable, bacteria react in different ways to survive. For instance, some of them produce endospore, with no active metabolism, which can survive for a very long time: this is level 0. If the conditions improve, the endospore germinates (back to level 1). Otherwise, they eventually die.

### 5.3.1    What If $M = 2$?

In this case, Equation 5.30 can be more conveniently rewritten as

$$\begin{cases} \dot{a}_1(t) = -i\omega_1 a_1(t) - i\lambda \left( a_0(t)\beta(t) + a_2(t)\beta^\dagger(t) \right), \\ \dot{a}_2(t) = -i\omega_2 a_2(t) - i\lambda a_1(t)\beta(t). \end{cases} \tag{5.32}$$

We have already noticed that $\beta(t)$ and $a_0(t)$ are no longer unknown of the problem here as they have been deduced previously. For instance, the first nontrivial contribution for $\beta(t)$ can be written as

$$\beta(t) \simeq \int_{\mathbb{R}} B_f(k) \, B_g^\dagger(k) \, e^{i(\Omega_g - \Omega_f)kt} \, dk.$$

With this approximation, after some computations, we deduce that

$$a_1(t) = a_1 e^{-i\omega_1 t} + \int_{\mathbb{R}} X_0(k)\mu_0(k,t) \, dk + \int_{\mathbb{R}} X_2(k)\mu_2(k,t) \, dk, \tag{5.33}$$

where we have defined the following functions

$$\begin{cases} \gamma_0(k) = i[(\Omega_g - \Omega_f)k - \omega_0] - \pi v^2, \\ \gamma_2(k) = -i[(\Omega_g - \Omega_f)k + \omega_2], \\ \mu_j(k,t) = \dfrac{1}{\gamma_j(k) + i\omega_1} \left( e^{\gamma_j(k)t} - e^{-i\omega_1 t} \right), \end{cases} \tag{5.34}$$

for $j = 0, 2$, and the following time-independent operators:

$$\begin{cases} X_0(k) = -i\lambda \, a_0 B_f(k) B_g^\dagger(k), \\ X_2(k) = -i\lambda \, a_2 B_f^\dagger(k) B_g(k). \end{cases} \tag{5.35}$$

**Remark:** The next step of our analysis is the computation of the mean value of the operator $\hat{n}_1(t) := a_1^\dagger(t) a_1(t)$. Because of Equation 5.33, this means that we need to compute, among other quantities, $\langle X_0^\dagger(k) X_0(q) \rangle$ and

$\langle X_2^{\dagger}(k)X_2(q)\rangle$. A simple-minded computation produces, as a consequence of Equation 5.20, with $n_f(k)$ and $n_g(k)$ replaced by $n_f$ and $n_g$,

$$\langle X_0^{\dagger}(k)X_0(q)\rangle = \lambda^2 n_0\, n_f(1+n_g)(\delta(k-q))^2,$$

and a similar result for $\langle X_2^{\dagger}(k)X_2(q)\rangle$. This is not a surprise; see Foot-note 3. Of course, such a result, which is due to the nonlinearity of the model described by our Hamiltonian (Eq. 5.1), needs a regularization. A possible way out is discussed in Bagarello (2008) and in references therein. In particular, it is shown that, with a suitable limiting procedure, the square of a Dirac delta is essentially nothing but a delta distribution itself. We import this idea here, therefore, replacing $(\delta(k-q))^2$ with $\delta(k-q)$.

With this in mind, calling $n_1(t) = \langle \hat{n}_1(t)\rangle$ and using complex contour integration again, we find that, if $\Omega_f \neq \Omega_g$,

$$n_1(t) = n_1 + \lambda^2 n_0\, n_f\, (1+n_g)\frac{1}{|\Omega_f - \Omega_g|}\left(1 - e^{-2\pi v^2 t}\right)$$

$$+ \lambda^2 n_2\, n_g(1+n_f)\frac{2\pi}{|\Omega_f - \Omega_g|}\, t. \tag{5.36}$$

Of course, owing to the fact that $\hat{N}$ in Equation 5.9 is a constant of motion, $n_2(t) = \langle \hat{n}_2(t)\rangle$ can be deduced simply using the results already found up to this stage, with not many further computations. In fact, as $\hat{N}_S(t) + \hat{N}_f(t) + \hat{N}_g(t) = \hat{N}_S(0) + \hat{N}_f(0) + \hat{N}_g(0)$, and as $\hat{N}_S(t) = \hat{n}_0(t) + \hat{n}_1(t) + \hat{n}_2(t)$, we have

$$\hat{n}_2(t) = N_S(0) + \hat{N}_f(0) + \hat{N}_g(0) - (\hat{N}_f(t) + \hat{N}_g(t) + \hat{n}_0(t) + \hat{n}_1(t)).$$

It is clear that the approximations we have considered so far make sense only for small values of $t$. This is because of the linear (in $t$) contribution in Equation 5.36, which necessarily breaks down the existence of the integral of motion $\hat{N}$. In fact, $n_1(t)$ becomes necessarily significantly large after some (large value of) time, so that it is impossible for $\hat{N}(t)$ to stay constant in time. We could use this argument to look for the largest value of time for which the approximation works fine: it is clear that for the value $\hat{t}$ of time for which $\langle \hat{N}(\hat{t})\rangle > \langle \hat{N}(0)\rangle$, the approximations introduced all along our computations do not work any longer.

Of course, the *disturbing* linear term in Equation 5.36 disappears when $n_2$ or $n_g$ are zero, that is, if the highest level of $\mathcal{S}$ or the reservoir of the garbage were initially empty.

**Remarks:**

1. Of course, a rigorous approach to the dynamical behavior of $\tilde{S}$ cannot avoid an estimate of the errors we have introduced in our treatment by adopting the approximations considered previously. However, as already stated several times, at the present stage of our research, we are much more interested in explicit results rather than on mathematical rigor. Rigor is just postponed, even if, not to have very bad surprises, all along this book we are also putting in evidence all the *dangerous steps*.

2. The case with $\Omega_f = \Omega_g$ cannot be considered as a special case of what we have discussed so far. This is clear from Equation 5.36, for instance. As it is not particularly interesting for us, this situation is not considered here.

3. It might be worth recalling that what we have considered in Section 5.2 is the *small $\lambda$ regime*, $\lambda \ll \min\left(\lambda_f, \lambda_g\right)$. The opposite regime, $\lambda \gg \max\left(\lambda_f, \lambda_g\right)$, looks much harder and an analytical solution cannot be found. Numerical techniques could be a good alternative. This work is in progress.

## 5.4  FINAL COMMENTS

In view of the many approximations that we have adopted to compute $n_1(t)$ and $n_2(t)$, we prefer to restrict the analysis of our results to the time dependence of $\hat{n}_0(t)$, $\hat{N}_f(t)$, and $\hat{N}_g(t)$ only. In particular, as already briefly discussed at the end of Section 5.2, we want to set up a test of efficiency of the system. Recalling that the limiting values for the averages of the above number operators are

$$\begin{cases} n_0^\infty = \dfrac{n_f\lambda_f^2\Omega_g + n_g\lambda_g^2\Omega_f}{\lambda_f^2\Omega_g + \lambda_g^2\Omega_f}, \\[3mm] n_f^\infty = n_f + n_0\,\dfrac{\lambda_f^2\Omega_g}{\lambda_f^2\Omega_g + \lambda_g^2\Omega_f}, \\[3mm] n_g^\infty = n_g + n_0\,\dfrac{\lambda_g^2\Omega_f}{\lambda_g^2\Omega_f + \lambda_f^2\Omega_{g]}}, \end{cases}$$

it looks natural to consider two opposite regimes. The first one corresponds to

$$\lambda_f^2\Omega_g \gg \lambda_g^2\Omega_f, \tag{5.37}$$

whereas the second regime is defined by the opposite inequality:

$$\lambda_f^2 \Omega_g \ll \lambda_g^2 \Omega_f. \tag{5.38}$$

It is interesting to notice that if Equation 5.37 is satisfied, then $n_0^\infty \simeq n_f$, $n_f^\infty \simeq n_f + n_0$, and $n_g^\infty \simeq n_g$. This means that the garbage does not change at all, whereas the food increases of the same amount of the entities, which at $t = 0$, were living in level 0. This is the best performance we can get for $\tilde{S}$: all the dying entities originally belonging to level 0 become food. No entity turns into garbage! Whenever the parameters of $H$ satisfy Equation 5.37, the system $\tilde{S}$ has an *optimal performance*.

On the other hand, let us suppose that the second inequality (Eq. 5.38) is satisfied. In this case, we find that $n_0^\infty \simeq n_g$, $n_f^\infty \simeq n_f$, and $n_g^\infty \simeq n_g + n_0$. As we can see, although the food is unchanged now, the garbage increases significantly: all the dying entities of level 0 now become garbage. No entity is transformed into food! Hence, when the parameters of $H$ satisfy Equation 5.38, $\tilde{S}$ has a very bad performance.

Concerning the speed of the decays of $n_0(t)$, $n_f(t)$, and $n_g(t)$, this is related to $\nu^2$, as we see from Equations 5.21, 5.24, and 5.27. The larger the value of $\nu^2$, the faster the decay of the functions to their limiting values; in particular in the optimal performance conditions, which we now rewrite as $\frac{\lambda_f^2}{\Omega_f} \ll \frac{\lambda_g^2}{\Omega_g}$, $\nu^2 \simeq \frac{\lambda_f^2}{\Omega_f}$. On the other hand, if Equation 5.38 is satisfied, then $\nu^2 \simeq \frac{\lambda_g^2}{\Omega_g}$. In both cases, the stronger the interaction with the reservoir, the faster the decay. In Figures 5.2–5.4, we plot the functions $n_0(t)$ (thin line), $n_f(t)$ (dashed line), and $n_g(t)$ (thick line), respectively, in the optimal performance regime ($\lambda_f = 0.1$, $\lambda_g = 0.01$, $\Omega_f = \Omega_g = 1$), in the opposite regime, that is, when Equation 5.38 is satisfied ($\lambda_g = 0.1$, $\lambda_f = 0.01$, $\Omega_f = \Omega_g = 1$), and in an intermediate regime ($\lambda_f = 0.1$, $\lambda_g = 0.07$, $\Omega_f = \Omega_g = 1$); fixing, in all these plots, the values are $n_0 = 10$, $n_f = 7$, and $n_g = 3$.

These figures show, in particular, that with our choice of the parameters and of the initial conditions, if Equation 5.37 is satisfied, then $n_g(t)$ stays constant in the time interval we have considered. Analogously, if Equation 5.38 holds true, $n_f(t)$ does not change with time.

**Remark:** It is quite interesting to observe that the performance of the system, as well as the speed with which some limiting values are obtained, are driven by the values of the interaction parameters $\lambda_f$ and $\lambda_g$, and also from the values of the free frequencies of the reservoirs, $\Omega_f$ and $\Omega_g$. Once again, we conclude that the free Hamiltonian is crucial in the analysis of the interacting system. However, we should also mention that no

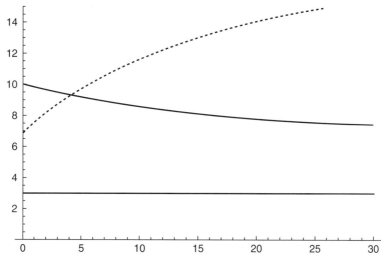

**Figure 5.2**    $n_0(t)$, thin line, $n_f(t)$, dashed line, and $n_g(t)$, thick line, when Equation 5.37 is satisfied.

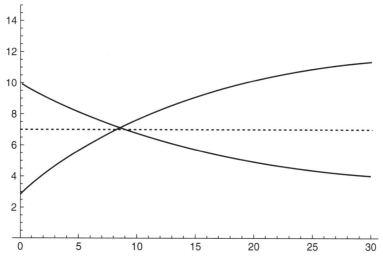

**Figure 5.3**    $n_0(t)$, thin line, $n_f(t)$, dashed line, and $n_g(t)$, thick line, when Equation 5.38 is satisfied.

dependence on the $\omega_l$s appears, suggesting that what is really important here is the role of the reservoirs. It is not hard to imagine that this could be simply related to our assumption that all the energy levels of $\mathcal{S}$ are equally spaced, or to the many approximations adopted to get an analytic expression for the solution.

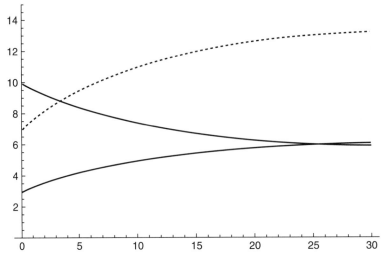

**Figure 5.4**  $n_0(t)$, thin line, $n_f(t)$, dashed line, and $n_g(t)$, thick line, in intermediate regime.

We conclude this section considering what happens in the following limiting situations:

**Situation 1:** Let us assume that Equation 5.37 holds, and that $n_f = 0$, $n_g \neq 0$.

In this case, $n_0^\infty \simeq 0$, $n_f^\infty \sim n_0$, and $n_g^\infty \simeq n_g$. This shows that all the entities originally occupying level 0 die after some time and $\mathcal{R}_f$, originally empty, is filled up with them. The garbage is not changed: there is a one-directional flux from level 0 to $\mathcal{R}_f$.

**Situation 2:** Let us assume that Equation 5.38 holds, and that $n_f \neq 0$, $n_g = 0$.

In this case, $n_0^\infty \simeq 0$, $n_f^\infty \simeq n_f$, and $n_g^\infty \simeq n_0$. This shows that all the entities originally occupying level 0 die after some time and $\mathcal{R}_g$, originally empty, is filled up with them. The food is not changed: there is a one-directional flux from level 0 to $\mathcal{R}_g$.

These results show, explicitly, that the initial conditions may *dynamically* produce a one-directional flow, even if the self-adjoint Hamiltonian originally describes two different flows, in opposite directions. Moreover, these results, as well as Figures 5.2 and 5.3, also suggest that $\mathcal{R}_f$ and $\mathcal{R}_g$ are, in a sense, absolutely equivalent in our analysis. What introduces a difference between the two is the ratio $\lambda_{f,g}^2/\Omega_{f,g}$, which describes the *biological meaning* of the reservoir.

# CHAPTER 6

# AN INTERLUDE: WRITING THE HAMILTONIAN

It is now time to stop for a moment and pay attention to the main ingredient of our approach, the Hamiltonian of the system $S$, which we use to deduce its time evolution. In particular, we consider separately two different physical situations, the ones corresponding to *closed* and to *open* systems, that is, those in which there is no interaction with any environment and those for which this kind of interaction plays a relevant role. The output of this chapter will be a set of rules, like the ones already considered in Chapter 3 and extending those discussed in Bagarello (2012), which can be used to write down the Hamiltonian of that particular classical system we are willing to describe.

## 6.1  CLOSED SYSTEMS

We begin with a system, $S$, made of *elements* $\tau_j$, $j = 1, 2, \ldots, N_S$, which are only interacting among themselves. This means, from the point of view of the interactions, that $S$ is the *entire world*: there is nothing outside. This is what we call here a *closed system*. Each element $\tau_j$ is defined by a certain set of variables (the self-adjoint *observable operators* for $\tau_j$): $(v_1^{(j)}, v_2^{(j)}, \ldots, v_{K_j}^{(j)})$. In other words, the $K_j$-dimensional vector

*Quantum Dynamics for Classical Systems: With Applications of the Number Operator,*
First Edition. Fabio Bagarello.
© 2013 John Wiley & Sons, Inc. Published 2013 by John Wiley & Sons, Inc.

$\mathbf{v}^{(j)} := (v_1^{(j)}, v_2^{(j)}, \ldots, v_{K_j}^{(j)})$ defines completely the dynamical status of $\tau_j$. To clarify what we mean, let $\tau_1$ be Bob in the love triangle considered in Section 3.3. Then, $v_1^{(1)}$ is Bob's LoA for Alice, $v_1^{(1)} = \hat{n}_{12} = a_{12}^\dagger a_{12}$, while $v_2^{(1)}$ is his LoA for Carla, $v_2^{(1)} = \hat{n}_{13} = a_{13}^\dagger a_{13}$. This means that Bob's status is completely described by a two-dimensional vector (here $K_1 = 2$), whose components are the two LoAs that Bob experiences for Alice and Carla: nothing more is required, here, to describe Bob, and for this reason, his dynamical behavior is entirely known when the time evolution of the vector $\mathbf{v}^{(1)} := (v_1^{(1)}, v_2^{(1)}) = (\hat{n}_{12}, \hat{n}_{13})$ is deduced. Let further $\tau_2$ be Alice. Her status is described by a one-dimensional vector, $\mathbf{v}^{(2)} := (v_1^{(2)}) = (\hat{n}_2)$, whose component is simply Alice's LoA for Bob. Then, $K_2 = 1$. Similarly, Carla is $\tau_3$, $K_3 = 1$, and her status is another one-dimensional vector, $\mathbf{v}^{(3)} := (v_1^{(3)}) = (\hat{n}_3)$, whose only component is Carla's LoA for Bob.

In the analysis of migrations (Section 4.3), $\mathcal{S}$ consists of just two elements, the first and the second species. The components of the vectors $\mathbf{v}^{(1)}$ and $\mathbf{v}^{(2)}$, in this case, are the local densities of the two species in the different lattice sites:

$$\mathbf{v}^{(1)} = \left(\hat{n}_1^{(a)}, \hat{n}_2^{(a)}, \ldots, \hat{n}_N^{(a)}\right), \qquad \mathbf{v}^{(2)} = \left(\hat{n}_1^{(b)}, \hat{n}_2^{(b)}, \ldots, \hat{n}_N^{(b)}\right).$$

Hence, $K_1 = K_2 = N$, which is the number of cells in the lattice.

Another example, which is considered in many details several times in Part II of this book, is provided by an SSM, our simplified stock market. Here $\mathcal{S}$ is made by $N_S$ traders $\tau_j$, $j = 1, 2, \ldots, N_S$, and the status of each one of these traders is defined by the number of shares and by the amount of money in their *portfolios*. If, for simplicity, we assume that only one kind of shares go around the SSM, then the vector $\mathbf{v}^{(j)}$ defining the status of $\tau_j$ is the following

$$\mathbf{v}^{(j)} = \left(\hat{n}_j, \hat{k}_j\right),$$

where $\hat{n}_j$ and $\hat{k}_j$ are the shares and the cash operators for the portfolio of $\tau_j$. Their mean values are useful to compute the explicit value of this portfolio.

In the language we are adopting here, describing the time evolution of $\mathcal{S}$ means, as already stated, giving the time evolution of (each) vector $\mathbf{v}^{(j)}$. Of course, we need to compute the time evolution of each operator-valued component of $\mathbf{v}^{(j)}$. Adopting a quantum-like description, this time evolution can be deduced by a suitable Hamiltonian, whose analytical expression is fixed by considering some guiding rules, which we discuss later.

The first natural requirement is what we call rule

**R1:** *In the absence of interactions between the various $\tau_j$s, all the vectors $\mathbf{v}^{(j)}$ stay constant in time.*

This is just our old **Rule 2**, stated in a more abstract and general language. In terms of the Hamiltonian of $\mathcal{S}$, **R1** means that if the interaction Hamiltonian is taken to be zero, then the (remaining) free Hamiltonian, $H_0$, must commute with the single operators appearing as components of the various $\mathbf{v}^{(j)}$, $v_i^{(j)}$, $j = 1, 2, \ldots, N_S$, $i = 1, 2, \ldots, K_j$. For instance, considering the love triangle earlier, this rule can be written as

$$[H_0, a_{12}^\dagger a_{12}] = [H_0, a_{13}^\dagger a_{13}] = [H_0, a_2^\dagger a_2] = [H_0, a_3^\dagger a_3] = 0.$$

In fact, if there is no interaction between Alice, Bob, and Carla, it is clear that there is no reason for their LoAs to change with time. This is a very general and natural rule for us: only the interactions cause a significant change in the status of the system. However, this does not prevent to have a nontrivial dynamics for other operators of $\mathcal{S}$, which are not observables, even in the absence of interactions. In fact, for instance, $[H_0, a_{12}] \neq 0$. Hence, it follows that $a_{12}(t) \neq a_{12}(0)$.

Let us now consider $\mathcal{S}$ *globally*, that is, paying attention not to its single element $\tau_j$ but to the whole set of the $\tau_j$s. Depending on the system we are considering, it might happen that, for some reason, some global function of the $\mathbf{v}^{(j)}$s is expected to stay constant in time. For instance, when dealing with the SSM, two such functions surely exist: the total number of shares and the total amount of cash. In fact, assuming that our market is closed, the cash and the shares can only be exchanged between the traders, although they cannot be created or destroyed. Considering Alice–Bob's (closed) love affair, the idea is that the sum of Bob's and Alice's LoAs stay constant in time. This is a simple way to state our *law of attraction*, introduced in Section 3.2. These examples are behind our second rule.

**R2:** *If a certain global function $f(\mathbf{v}^{(1)}, \mathbf{v}^{(2)}, \ldots, \mathbf{v}^{(N_S)})$ is expected to stay constant in time, then $H$ must commute with $f$: $[H, f(\mathbf{v}^{(1)}, \ldots, \mathbf{v}^{(N_S)})] = 0$.*

Note that here, contrary to what we have done in **R1**, we are working with the full Hamiltonian $H$ and not just with $H_0$: in fact, here, the interactions are the interesting part of the problem. For instance, in the simple love affair described in Section 3.2, the Hamiltonian of the interacting system, $H$, should commute with $a_1^\dagger a_1 + a_2^\dagger a_2$, which represents the *total love of the system*. On the other hand, $[H, a_1^\dagger a_1]$ and $[H, a_2^\dagger a_2]$ are both different from zero: the single LoA changes, but the sum of the two does not. Analogously, if $c_j$ is the annihilation operator associated with the cash in the portfolio of the trader $\tau_j$, see Section 8.1, then, calling $\hat{K} = \sum_{j=1}^{N_S} c_j^\dagger c_j$ the *global cash operator*, $H$ must commute with $\hat{K}$: $[H, \hat{K}] = 0$.

**R2** can be generalized to more constants of motion. For instance, $H$ must commute also with $\hat{N} = \sum_{j=1}^{N_S} a_j^\dagger a_j$, where $a_j$ is the annihilation operator of the shares in $\tau_j$s portfolio. $\hat{N}$ is the *global shares operator*.

The third rule we need to introduce is again concerned with the possible interactions between the different elements of $S$. In these notes, we only consider a *two-body interaction*, that is, the possibility of one element of $S$ to interact with a single other element in a given instant of time. But this does not imply that $\tau_1$ cannot interact with both $\tau_2$ and $\tau_3$. It only means that this cannot happen simultaneously: $\tau_1$ interacts first with $\tau_2$ and, only after this interaction has taken place, with $\tau_3$.

Suppose now that $\tau_1$ and $\tau_2$ interact and that, because of this interaction, they change their status. For instance, the initial vector $<\mathbf{v}^{(1)}(b.i.)>$ after the interaction is replaced by $<\mathbf{v}^{(1)}(a.i.)>$. Here, $b.i.$ and $a.i.$ stand for *before* and *after interaction*, respectively, and $<\mathbf{v}>$ means that we are considering the mean value of the operator-valued vector $\mathbf{v}$ on a suitable state. Analogously, $<\mathbf{v}^{(2)}(b.i.)>$ changes to $<\mathbf{v}^{(2)}(a.i.)>$. To be concrete, we suppose here that the vectors $<\mathbf{v}^{(1)}(a.i.)>$ and $<\mathbf{v}^{(1)}(b.i.)>$ differ only for the values of their first two components, $<v_1^{(1)}>$ and $<v_2^{(1)}>$, and we call $\delta v_j^{(1)} = <v_j^{(1)}(a.i.)> - <v_j^{(1)}(b.i)>$, $j = 1, 2$, the differences between these values. We call $a_1^{(1)}$ and $a_2^{(1)}$ the two (bosonic or fermionic) annihilation operators associated with these components. Analogously, let us suppose that $<\mathbf{v}^{(2)}(a.i.)>$ and $<\mathbf{v}^{(2)}(b.i.)>$ also differ only for the values of their first two components, $<v_1^{(2)}>$ and $<v_2^{(2)}>$, and let $a_1^{(2)}$ and $a_2^{(2)}$ be the related annihilation operators. Then, we introduce $\delta v_j^{(2)} = <v_j^{(2)}(a.i.)> - <v_j^{(2)}(b.i)>$, $j = 1, 2$. To build up the interaction Hamiltonian $H_{\text{int}}$, which is responsible for such a change, we first have to consider the signs of the various $\delta v_j^{(k)}$. To fix the ideas, we suppose here that $\delta v_1^{(1)}$ and $\delta v_2^{(2)}$ are positive, while $\delta v_2^{(1)}$ and $\delta v_1^{(2)}$ are negative. Then our third rule can be stated as in the following.

**R3:** *The interaction Hamiltonian $H_{int}$ responsible for the earlier changes must contain the contribution*

$$\left(a_1^{(1)\,\dagger}\right)^{\delta v_1^{(1)}}\left(a_2^{(1)}\right)^{\delta v_2^{(1)}}\left(a_1^{(2)}\right)^{\delta v_1^{(2)}}\left(a_2^{(2)\,\dagger}\right)^{\delta v_2^{(2)}}, \tag{6.1}$$

*together with its Hermitian conjugate.*

The need to introduce the Hermitian conjugate in our Hamiltonians was already discussed several times. The point is that this is the easiest way to introduce a unitary evolution, and this is crucial if we want to use the Heisenberg equation of motion (Eq. 2.2). From Equation 6.1, we observe that creation operators appear for positive values of $\delta v_j^{(k)}$'s, while annihilation operators are used for negative $\delta v_j^{(k)}$'s. The reason is clear: as, for instance, $\delta v_1^{(1)}$ is positive, during the interaction between $\tau_1$ and $\tau_2$ the value of $<v_1^{(1)}>$ increases, and this increment can only be produced (in our settings) by creation operators. On the other hand, as $\delta v_1^{(2)}$ is negative, the value of $<v_1^{(2)}>$ decreases, and this can be well described by considering an annihilation operator in $H_{int}$. It is not particularly difficult to extend Equation 6.1 to slightly different situations, for instance, when more components of the status vectors are involved in the interaction between $\tau_1$ and $\tau_2$. It is also very simple to go from this, in general, nonlinear interaction Hamiltonian to its linear version: this is what happens when all the variation parameters $\delta v_j^{(k)}$ are equal to 1. The Hamiltonian in Equation 3.5 is the first example of how **R3** is implemented in a simple situation, where the nonlinearity is *related* to a single variable and to a single actor, Bob. In Section 3.2, we have also discussed why this nonlinearity could be considered a measure of the *relative reactions* of the two actors of the game. Mutatis mutandis, the same **R3** can be used to deduce the Hamiltonian in Equation 3.14, as well as the one in Equation 4.3 or that introduced in Section 4.3, for instance. In Part II of this book, more Hamiltonians for closed systems are constructed, following **R1**, **R2**, and **R3** and some minor modifications of these rules. Quite obviously, the difficulties of the dynamics deduced by these Hamiltonians will be directly proportional to the complexity of the system under consideration.

## 6.2  OPEN SYSTEMS

As discussed several times, looking at a certain classical system $S$ as if it were a closed system, it might not necessarily be the best point of view.

In fact, quite often, the role of the environment turns out to be crucial in producing a reasonable dynamical behavior for $\mathcal{S}$. For instance, going back to Chapter 3, it is only the presence of a reservoir that produces some damping in the time evolution of Alice's and Bob's LoAs, which, only in this way, can reach some equilibrium in their love affair. In the absence of such a reservoir, as we have already discussed, more likely, oscillations take place and no equilibrium is reached at all! Hence, the outer world is what is needed to get a stable sentimental relationship. As in the standard literature, an open system for us is simply a system $\mathcal{S}$ interacting with a reservoir $\mathcal{R}$. From a dynamical point of view, the Hamiltonian $H_{\tilde{\mathcal{S}}}$ of this larger system, $\tilde{\mathcal{S}} := \mathcal{S} \cup \mathcal{R}$, appears to be the Hamiltonian of an interacting system. This means that the general expression of $H_{\tilde{\mathcal{S}}}$ is the sum of three contributions:

$$H_{\tilde{\mathcal{S}}} = H_{\mathcal{S}} + H_{0,\mathcal{R}} + H_{\mathcal{S},\mathcal{R}}. \tag{6.2}$$

Here $H_{\mathcal{S}}$ is the Hamiltonian for $\mathcal{S}$, the explicit expression of which should be deduced adopting **R1**, **R2**, and **R3** of the previous section, working as if $\mathcal{S}$ was a closed system. This implies, among other things, that $H_{\mathcal{S}}$ contains a free contribution plus a second term describing the interactions of the various elements of $\mathcal{S}$ among themselves. $H_{0,\mathcal{R}}$ is the free Hamiltonian of the reservoir, while $H_{\mathcal{S},\mathcal{R}}$ contains the interactions between the system and the reservoir. To fix the explicit expression of $H_{0,\mathcal{R}}$, we follow the following rule, which extends to the reservoir **R1**:

**R4:** *In the absence of any interaction, $H_{0,\mathcal{R}}$ must not change the status of the observables of the reservoir.*

In the models discussed in this book, the only relevant *observables of the reservoir* are the number operators associated with it. For instance, in Chapter 3, there were two such operators, $\int_{\mathbb{R}} A^\dagger(k) A(k)\, dk$ and $\int_{\mathbb{R}} B^\dagger(k) B(k)\, dk$, both commuting with the Hamiltonian $H$ in Equation 3.20 in the absence of interactions, that is, if we fix $\gamma_A = \gamma_B = \lambda = 0$. Analogously, in Chapter 5, the number operators $\hat{N}_f$ and $\hat{N}_g$ for the reservoirs defined in Equation 5.4 commute with the free Hamiltonian, which is given by Equations 5.1 and 5.3. In this case, it is clear that **R4** is satisfied.

**Remark:** Of course, we could introduce a dynamical, operator-valued, vector **V** also for the reservoir, in analogy with what we have done for the system. However, because of the previous remark on the observables of the reservoir, we feel there is no need for this generalization here.

The last ingredient in Equation 6.2 is $H_{S,R}$. Its analytic expression is fixed by the simplest possible requirement, which can be seen as a simple extension of **R3** and is given by the following rule.

**R5:** *In the interaction Hamiltonian $H_{S,R}$, each annihilation operator of the system is linearly coupled to a corresponding creation operator of the reservoir as in Equation 6.1, and vice versa.*

For instance, in the damped love affair described in Section 3.4, Alice's annihilation and creation operators are linearly coupled to the creation and annihilation operators of Alice's reservoir: the Hamiltonian contains the contribution $\int_{\mathbb{R}} \left( a^\dagger A(k) + a\, A^\dagger(k) \right) \, dk$. This same coupling is also assumed between Bob and his reservoir: $\int_{\mathbb{R}} \left( b^\dagger B(k) + b\, B^\dagger(k) \right) \, dk$. A similar expression is considered also in Section 4.4, in each cell of the lattice, when considering the migration of the two species. Moreover, the analysis developed in Chapter 5, is entirely based on this kind of Hamiltonian (Eq. 5.6). One might wonder why this particularly simple form of the interaction is assumed. The reason is very simple: because it works! What we need here is to produce damping, and this simple choice of $H_{S,R}$ indeed produces damping, with not many additional analytical complications. However, it is worth noticing that when we deal with an open system, the Hamiltonian is only part of the treatment (a big part, by the way!). We also require some *extra care* while fixing the initial conditions, because a smart choice of these can produce, out of a self-adjoint (and, therefore, bidirectional)[1] Hamiltonian, a flux flowing in only one direction. This is what we wanted to describe, for instance, in Chapter 5, where we were interested in a one-directional flux from $\mathcal{R}_f$ to the levels $1, 2, \ldots M$, while a two-directional flux was allowed only between level 0 and $\mathcal{R}_f$.

It is not hard to see that the five rules discussed in this chapter do not cover all the situations considered in this book. For instance, the jump Hamiltonian (Eq. 5.8) does not strictly obey **R5**. However, it obeys to an extension of this rule, extension that was physically motivated before. Another extension of **R3** is the one giving rise to $\tilde{H}_I$ in Equation 8.1. We postpone the analysis of this extension to Chapter 8, which describes a *selling* and *buying* procedure between traders.

---

[1]Due to the fact that $H$ must be self-adjoint, if it contains, for example, a term $a_1^\dagger a_2$, it must also contain $a_2^\dagger a_1$. They describe, of course, opposite *movements*. For this reason, we speak of bidirectionality.

## 6.3 GENERALIZATIONS

We have considered in Section 2.7.1 the possibility of using some *effective Hamiltonian* in the attempt to describe damping effects. This procedure works efficiently in quantum optics, but as we have shown explicitly, does not seem applicable in this context without some deeper analysis. This is the main reason why we prefer to make use of some reservoir interacting with the system we want to describe, following the general ideas described first in Section 2.7 and adopting the rules proposed in this chapter. However, a second reason also exists, and this is maybe more important: what it might appear, at a first reading, just as a formal object, our *quantum reservoir*, should be really considered as the *true*, macroscopic environment surrounding the system and interacting with it. From this point of view, its presence is not only necessary but also very natural as we want to describe realistic situations rather than oversimplified dynamical systems.

What is quite interesting, and looks like a worthy extension of the framework considered so far, is the possibility of using some explicitly time-dependent Hamiltonian, $H(t)$. This may have interesting implications in the analysis of some systems. For instance, going back to the diffusion coefficients $p_{\alpha,\beta}$ introduced in Formula 4.10, it is surely interesting to admit some time-dependence to consider the possibility that not all the cells of the lattice maintain their conditions stable in time: if, at $t = 0$, the species $S_1$ may find some convenience in going from the cell $\alpha$ to the cell $\beta$, this might not be true at some later time, of course. Also, replacing the mobility coefficients $\gamma_a$ and $\gamma_b$ with two time-dependent real functions, $\gamma_a(t)$ and $\gamma_b(t)$, is surely an interesting extension of the model: the species might move slowly or fast depending on many possible factors. Going from a Hamiltonian with no explicit dependence on time to a Hamiltonian with such a dependence changes the rules of the game quite a bit. In particular, Equation 2.2 cannot be used, and it might be more convenient to adopt some perturbative approach like the one sketched in Section 2.5.1 and adopted in Chapter 9.

Considering time-dependent Hamiltonians might be useful also to add some stochasticity to the system. This could be particularly interesting, for instance, in the analysis of the SSMs. These are aspects that deserve further investigations, and we hope to be able to consider these aspects soon.

Another kind of possible generalization is related to the use of non-commuting observables in the definition of the physical system. This is suggested in many papers; see Segal and Segal (1998), for instance, where quantum mechanical ideas are imported in an economical context exactly because of the existence of incompatible observables: in the work by

Segal and Segal (1998), these observables are the price at time $t$ and its time derivative at the same time. These two quantities cannot be both known accurately as otherwise each trader possessing both these information would earn as much money as he likes. The impossibility of knowing the price and its time derivative was considered by these authors as the evidence of some intrinsic Heisenberg uncertainty principle; see Section 2.4. So far, in our research, we have not considered this aspect of the quantum world, but we plan to go back to this in the future. In particular, for instance, the order of the operators appearing in Equation 6.1 is not relevant for us here as all the operators involved do commute (or anticommute).[2]

---

[2] In this case, an overall sign in general appears.

# PART II

# SYSTEMS WITH MANY ACTORS

# CHAPTER 7

# A FIRST LOOK AT STOCK MARKETS

The applications we have considered so far involve a finite, and usually small, number of actors (or sets of actors), two or three at most: Alice, Bob, and Carla in Chapter 3, or species $S_1$ and $S_2$ in Chapter 4. In Chapter 5, we have considered a single group of entities, $S$ (e.g., bacteria). However, our analysis clearly shows that this does not mean that the corresponding systems have just two or three degrees of freedom. For instance, the description of $S_1$ and $S_2$ in the lattice $\mathcal{R}$ requires the use of $2N$ independent fermionic operators, two for each lattice site. Analogously, the independent variables needed to describe $S$ in Chapter 5 are $N + 1$, one for each LoW of the system. Moreover, as clarified in Section 2.7, coupling an external reservoir to the system makes the number of degrees of freedom suddenly jump to infinity. For this reason, the models considered in Part I cannot be really seen as *simple*–they are not simple at all! However, they do not involve so many different *ingredients*.

In this part of the book, we consider more *complicated* systems, that is, those systems with *genuinely* many actors involved. In particular, we concentrate on *simplified stock markets* (SSMs) of the kind we have already briefly introduced in these notes. Our interest is directed to the *portfolio* of the various actors of the market, the so-called *traders*, who buy or sell *shares* of some kind, paying for that the value of the share at the time

*Quantum Dynamics for Classical Systems: With Applications of the Number Operator*,
First Edition. Fabio Bagarello.
© 2013 John Wiley & Sons, Inc. Published 2013 by John Wiley & Sons, Inc.

when transaction takes place. The portfolio is based on two main ingredients: the cash and the shares. At a given time, its value is simply the sum of the total amount of cash and the total value of the shares belonging to the particular trader we are considering. Later, in Chapter 9, we also consider a different problem, looking for a certain *transition probability* between two different configurations of the market. Other than cash and shares, two more actors will behave as dynamical variables: the price of the shares and what we will call the *market supply*, a sort of *global tendency* of the traders to buy or sell shares.

In particular, in this chapter, we propose some motivations suggesting the use of the number representation in an extremely simplified stock market. After that, we introduce and discuss in detail the first model that has a very important technical feature: the differential equations of motion deduced adopting this model can be explicitly solved. Therefore, this solution gives us a first hint of what the temporal evolution of the single trader may look like in our settings.

## 7.1   AN INTRODUCTORY MODEL

The model we discuss in this section is really an oversimplified toy model of a closed stock market (i.e., not interacting with any reservoir) based on the following assumptions:

1. Our market consists of $L$ traders exchanging a single kind of share.
2. The total number of shares, $N$, is fixed in time.
3. A trader can only interact with a single other trader, that is, the traders feel only a *two-body interaction*.
4. The traders can buy or sell only one share in each single transaction.
5. There exists a *unique* price for the share, fixed by the market. In other words, we are not considering any difference between the price at which the share is *offered to the other traders*, the *ask price*, and the price that the trader is willing to pay for the share, the *bid price*.
6. The price of the share changes discontinuously, with discrete steps, multiples of a given *monetary unit*.
7. Each trader has a *huge* amount of money that can be used to buy the shares.

Let us briefly comment on the assumptions, most of which are meant to produce a reasonably simple model to be handled analytically. Assuming that there is only a single kind of share may appear rather restrictive, but, as we will see later, more kinds of shares can be introduced without changing the strategy; the only price to pay is an increasing difficulty in the analytic computations. The third assumption given earlier simply means that it is not possible for, say, the traders $\tau_1$, $\tau_2$, and $\tau_3$ to interact with each other at the same time. However, $\tau_1$ can interact directly with $\tau_3$ or via its interaction with $\tau_2$. In other words, $\tau_1$ interacts first with $\tau_2$ and only in the second step, $\tau_2$ interacts with $\tau_3$. This is a typical simplification considered in Many-Body Theory in which, quite often, all the $N$-body interactions, $N \geq 3$, are assumed to be negligible with respect to the two-body interaction. This can be understood simply because the probability of having simultaneously an interaction involving three or more traders is very small, and surely smaller than the probability that only two traders interact. Assumptions 4, 5, and 7 are useful to simplify the model and to allow us to extract some driving ideas that are used to build up more realistic descriptions. In particular, Assumption 7 means that there is no reason to consider, here, the cash as a dynamical variable; only the number of shares needs to be considered to somehow fix the value of the portfolios. Assumption 6 just reflects the fact that, for instance, there is nothing less than 1 cent of euro, whereas Assumption 2 is a consequence of the fact that the market is closed. Some of these assumptions are relaxed and made more realistic in the next chapter. Of course, as *repetita juvant*, we want to remark once again that the model we are going to construct using these simple recipes is very far from being a realistic stock market. However, we believe that it is already a reasonably good starting point, useful to clarify some features of our approach and to suggest more realistic extensions.

As usual, the time behavior of this model can be described by the *Hamiltonian* of the model, which describes the free evolution of the model plus the effects of the interaction between the traders. The Hamiltonian of this simple model is taken to be the following:

$$H = H_0 + H_{\text{int}}, \quad H_0 = \sum_{l=1}^{L} \alpha_l a_l^\dagger a_l + \epsilon p^\dagger p, \quad H_{\text{int}} = \sum_{i,j=1}^{L} p_{ij} a_i a_j^\dagger,$$

$$(7.1)$$

where the following CCR are assumed:

$$[a_l, a_n^\dagger] = \delta_{ln}\, \mathbb{1}, \, [p, p^\dagger] = \mathbb{1},$$ $$(7.2)$$

while all the other commutators are zero. The meaning of these operators is easily deduced by the general ideas introduced in Section 2.2: $a_l$ and $a_l^\dagger$, respectively, destroys and creates a share in the portfolio of $\tau_l$, whereas the operators $p$ and $p^\dagger$ modify the price of the share (down and up).

The coefficient $p_{ij}$ in Equation 7.1 takes the value 1 or 0, depending on whether $\tau_i$ interacts with $\tau_j$. We also assume that $p_{ii} = 0$ for all $i$, which simply means that $\tau_i$ does not interact with himself or herself.

Concerning $H_0$ in Equation 7.1, this is chosen adopting **Rule 2** of Section 3.4 or **R1** of Chapter 6: whenever there is no interaction between the traders ($p_{ij} = 0$ for all $i$ and $j$), all the number operators associated with our SSM commute with $H_0$; see later discussion. On the other hand, the interaction Hamiltonian $H_{int}$ obeys a linear version of **R3**: more explicitly, the contribution $a_i\, a_j^\dagger$ in $H_{int}$ *destroys* a share belonging to the trader $\tau_i$ and *creates* a share in the portfolio of the trader $\tau_j$. In other words, if $p_{ij} = 1$, the trader $\tau_i$ sells a share to $\tau_j$. However, as $H$ must be self-adjoint, $p_{ij} = 1$ also implies that $p_{ji} = 1$. This means that the interaction Hamiltonian contains both the possibility that $\tau_i$ sells a share to $\tau_j$ and the possibility that $\tau_j$ sells a share to $\tau_i$. Different values of $\alpha_j$ in the free Hamiltonian are then used, in principle, to introduce a sort of *ability* of the trader, which will make more likely that the *most expert* trader conveniently sells or buys his shares to or from the other traders to increase the value of his portfolio as much as he can.

The only observables whose time evolution we are interested in are, clearly, the price of the share and the number of shares of each trader. In fact, as we have already remarked, within our simplified scheme there is no room for the cash! The *price operator* is $\hat{P} = \epsilon p^\dagger\, p$, while the *j-number operator* is $\hat{n}_j = a_j^\dagger\, a_j$, which represents the number of shares that $\tau_j$ possesses. The operator describing the *total number of shares* is the sum of several contributions: $\hat{N} = \sum_{j=1}^{L} \hat{n}_j = \sum_{j=1}^{L} a_j^\dagger\, a_j$. The choice of $H$ in Equation 7.1 is also suggested by the requirement 2 given earlier, which reflects **R2** of Chapter 6. Indeed, it is easy to check, using Equation 7.2, that $[H, \hat{N}] = 0$. This implies that the time evolution of $\hat{N}$, $\hat{N}(t) = e^{iHt}\hat{N}\, e^{-iHt}$ is trivial: $\hat{N}(t) = \hat{N}(0)$ for all $t$. However, as $[H, \hat{n}_j] \neq 0$, $\hat{n}_j(t)$ surely depends on time. This is clear from the definition of $H$: the term $\sum_{l=1}^{L} \alpha_l\, a_l^\dagger a_l$ does not change the number of shares of the different traders, but only counts this number. On the contrary, the interaction $\sum_{i,j=1}^{L} p_{ij} a_i a_j^\dagger$ destroys a share belonging to $\tau_i$ but, at the same time, creates another share in the portfolio of the trader $\tau_j$. In doing this, the number of the shares of the single trader is changed, although their total number remains the same. It may be worth noticing that if all the $p_{ij}$ are zero, that is, if there is no interaction between the traders, then we have $H \equiv H_0$ and we get, consequently, $[H, \hat{n}_j] = 0$: our model produces an essentially stationary market, as it is expected. Therefore, **R1** of Chapter 6 is satisfied, as already stated.

We have implemented Assumptions 5 and 6 by requiring that the price operator $\hat{P}$ has the form given earlier, $\hat{P} = \epsilon p^\dagger p$, where $\epsilon$ is the monetary unit. Such an operator is assumed to be part of $H$; see Equation 7.1. In addition, because of the simplifications that have been considered in this toy model, $\hat{P}$ is clearly a constant of motion: $[H, \hat{P}] = 0$. This is not a realistic assumption and is relaxed in the subsequent chapters. However, it is used here because it provides a better understanding of the meaning of $\alpha_l$, as we see later.

We now need to describe the *state of the system*. In particular, we assume that, at time $t = 0$, the portfolio of the first trader is made by $n_1$ shares, the one of $\tau_2$ by $n_2$, and so on, and that the price of the share is $P = M\epsilon$. This means that the market is in a vector state $\omega_{n_1,n_2,\dots,n_L;M}$ defined by the following vector of the Hilbert space $\mathcal{H}$

$$\varphi_{n_1,n_2,\dots,n_L;M} := \frac{1}{\sqrt{n_1! n_2! \dots n_L! M!}} (a_1^\dagger)^{n_1} (a_2^\dagger)^{n_2} \cdots (a_L^\dagger)^{n_L} (p^\dagger)^M \varphi_0.$$
(7.3)

Here $\varphi_0$ is the *vacuum* of the model $a_j \varphi_0 = p \varphi_0 = 0$ for all $j = 1, 2, \dots, L$. If $B \in \mathcal{A}$, $\mathcal{A}$ being the *algebra* of the observables of our market, then

$$\omega_{n_1,n_2,\dots,n_L;M}(B) = < \varphi_{n_1,n_2,\dots,n_L;M}, B \varphi_{n_1,n_2,\dots,n_L;M} >,$$
(7.4)

where $<, >$ is the scalar product in $\mathcal{H}$. The Heisenberg equations of motion (Eq. 2.2) for the annihilation operators $a_l(t)$ produce the following very simple differential equation:

$$i\dot{a}(t) = Xa(t),$$
(7.5)

where we have introduced the matrix $X$, independent of time, and the vector of the unknowns, $a(t)$, as follows:

$$X \equiv \begin{pmatrix} \alpha_1 & p_{21} & p_{31} & \cdot & \cdot & p_{L-11} & p_{L1} \\ p_{12} & \alpha_2 & p_{32} & \cdot & \cdot & \cdot & p_{L2} \\ p_{13} & p_{23} & \alpha_3 & \cdot & \cdot & \cdot & \cdot \\ \cdot & \cdot & \cdot & \cdot & \cdot & \cdot & \cdot \\ \cdot & \cdot & \cdot & \cdot & \cdot & \cdot & \cdot \\ p_{1L-1} & p_{2L-1} & p_{3L-1} & \cdot & \cdot & \alpha_{L-1} & p_{LL-1} \\ p_{1L} & p_{2L} & p_{3L} & \cdot & \cdot & p_{L-1L} & \alpha_L \end{pmatrix},$$

$$
a(t) \equiv \begin{pmatrix} a_1(t) \\ a_2(t) \\ a_3(t) \\ \cdot \\ \cdot \\ a_{L-1}(t) \\ a_L(t) \end{pmatrix}.
$$

Notice that owing to the conditions on $p_{ij}$, and as all the $\alpha_l$ values are real, the matrix $X$ is self-adjoint. Equation 7.5 can now be solved as follows: let $V$ be the (unitary) matrix that diagonalizes $X$: $V^\dagger X V = diag\{x_1, x_2, \ldots, x_L\} =: X_d$, $x_j$ being its eigenvalues, $j = 1, 2, \ldots, L$. Matrix $V$ does not depend on time. Then, putting

$$
U(t) = \begin{pmatrix} e^{ix_1 t} & 0 & 0 & \cdot & \cdot & 0 \\ 0 & e^{ix_2 t} & 0 & \cdot & \cdot & 0 \\ 0 & 0 & e^{ix_3 t} & \cdot & \cdot & 0 \\ \cdot & \cdot & \cdot & \cdot & \cdot & \cdot \\ \cdot & \cdot & \cdot & \cdot & \cdot & \cdot \\ 0 & 0 & 0 & \cdot & \cdot & e^{ix_L t} \end{pmatrix},
$$

we get

$$
a(t) = V U(t) V^\dagger a(0), \tag{7.6}
$$

where, as it is clear, $a(0)^T = (a_1, a_2, \ldots, a_L)$. If we further introduce the adjoint of the vector $a(t)$, $a^\dagger(t) = (a_1^\dagger(t), a_2^\dagger(t), \ldots, a_L^\dagger(t)) = a^\dagger(0) V U^\dagger(t) V^\dagger$, we can explicitly check that $\hat{N}$ is a constant of motion. Indeed, we have

$$
\hat{N}(t) = a_1^\dagger(t) a_1(t) + a_2^\dagger(t) a_2(t) + \ldots + a_L^\dagger(t) a_L(t) = a^\dagger(t) \cdot a(t)
$$
$$
= (a^\dagger(0) V U^\dagger(t) V^\dagger) \cdot (V U(t) V^\dagger a(0)) = a^\dagger(0) \cdot a(0) = \hat{N}(0),
$$

where the dot denotes the usual scalar product in $\mathbb{C}^L$.

In order to analyze the time behavior of the different $\hat{n}_j(t)$, we simply have to compute the mean value $n_j(t) = \omega_{n_1, n_2, \ldots, n_L; M}(\hat{n}_j(t))$. As we have seen, this means that for $t = 0$, the first trader owns $n_1$ shares, the second trader, $n_2$, and so on, and that the price of the share is $M\epsilon$. It should be mentioned that the only way in which a matrix element such as $\omega_{n_1, n_2, \ldots, n_L; M}(a_j^k (a_l^\dagger)^m)$ can be different from zero is when $j = l$ and $k = m$. This follows from the orthonormality of the set $\{\varphi_{n_1, n_2, \ldots, n_L; M}\}$, which, as discussed in Section 2.1, is a direct consequence of the CCR, and from the action of the annihilation and creation operators on the vectors $\varphi_{n_1,n_2,\dots,n_L;M}$.

The easiest way to get the analytic expression for $n_j(t)$ is to fix the number of traders giving rise to our SSM, starting with the simplest situation, $L = 2$. In this case, we find that

$$
\begin{cases}
n_1(t) = \dfrac{1}{\Omega^2}\left\{n_1\left(\alpha^2 + 2\tilde{p}^2(1+\cos(\Omega t))\right) + 2\tilde{p}^2 n_2\left(1-\cos(\Omega t)\right)\right\}, \\[2mm]
n_2(t) = \dfrac{2\tilde{p}^2 n_1}{\Omega^2}\left(1-\cos(\Omega t)\right)) + n_2\left(1 + \dfrac{2\tilde{p}^2}{\Omega^2}\left(\cos(\Omega t)-1\right)\right),
\end{cases}
\tag{7.7}
$$

where we have defined $\Omega^2 = \alpha^2 + 4\tilde{p}^2$, with $\alpha = \alpha_2 - \alpha_1$ and $\tilde{p} = p_{12} = p_{21}$.

This result is not very different from the one given in Equations 4.6 and 4.7 because of the simplifying assumptions considered here and there. Hence, it is not surprising that some of the conclusions are similar to those deduced in Chapter 4. For instance, it is not hard to check that $n_1(t) + n_2(t) = n_1 + n_2$, as expected. In addition, if $\tilde{p} = 0$, we see that $n_1(t) = n_1$ and $n_2(t) = n_2$ for all $t$. This is natural and expected because when $\tilde{p} = 0$, there is no interaction at all between the traders, so that there is no reason for $n_1(t)$ and $n_2(t)$ to change in time. This is an explicit realization of rule **R1**. Another interesting consequence of Equation 7.7 is that, if $n_1 = n_2 = n$, that is, if the two traders begin interacting having the same number of shares, they do not change this equilibrium during the time; this is because the solution in Equation 7.7, under this condition, produces $n_1(t) = n_2(t) = n$. Again, this result is not particularly surprising. As both $\tau_1$ and $\tau_2$ possess the same amount of money (their huge sources!) and the same number of shares, why should they try to change their *safe* equilibrium? The role of $\alpha_1$ and $\alpha_2$, in this case, is unessential. In more general situations, it is clear that each $n_j(t)$ is a periodic function whose period, $T = \frac{2\pi}{\Omega}$, decreases for $|\alpha| = |\alpha_1 - \alpha_2|$ and $\tilde{p}$ increasing. Finally, if we call $\Delta n_j = \max_{t\in[0,T]}|n_j(t) - n_j(0)|$, which represents the highest variation of $n_j(t)$ in a period, we can easily deduce that $\Delta n_1 = \Delta n_2 =: \Delta_n = \frac{4\tilde{p}^2}{(\alpha_2-\alpha_1)^2+4\tilde{p}^2}|n_1 - n_2|$. As in Section 4.2, we see that $\Delta n$ tends to zero when $|\alpha_2 - \alpha_1|$ tends to infinity, while, if $\alpha_2 \simeq \alpha_1$, then $\Delta n$ almost coincides with $|n_1 - n_2|$, which is simply the absolute value of the difference between the initial number of shares of the two traders. Therefore, as in Section 4.2, it is natural to interpret $|\alpha_2 - \alpha_1|$ as the *inertia* of the system because when this quantity increases, the portfolios of the traders behave almost statically.

**Remarks:** (1) Recalling that, in principle, the number of shares should be integer, while $n_1(t)$ and $n_2(t)$ are continuous functions in $t$, we could introduce a *time for the m-th transaction*, $T_m$, chosen in such a way that $n_j(T_1)$, $n_j(T_2)$, ... are all integers, $j = 1, 2$. More easily, perhaps, and in agreement with what we have already discussed in Chapter 3, we could simply consider these continuous functions, and the related plots, as giving an indication of the behavior of the related discrete quantities: if the function $n_j(t)$ increases in the time interval $[t_a, t_b]$, then $\tau_j$ is buying shares, whereas he is selling shares if $n_j(t)$ decreases.

(2) Not surprisingly, the analytic time evolution is periodic and does not allow any decay. This makes sense as our market is closed (the total number of shares is preserved): because no reservoir is considered in the model, and as $H$ is self-adjoint, there is no easy way to produce any damping (Section 2.7).

(3) In the subsequent chapters, rather than $n_j(t)$, we are interested in the *portfolio* of the various traders, which, as stated, is nothing but the sum of the cash and the value of the shares of each single trader. In the present model, in view of the unessential role of the cash, the portfolio and the number of shares give exactly the same information. This is why we have considered here only the time evolution of the number of shares.

Let us now consider a market with three traders. In Figure 7.1, we plot $n_3(t)$ with the initial conditions $n_1 = 40$, $n_2 = n_3 = 0$, with $p_{12} = p_{13} = p_{23} = 1$ and different values of $\alpha_1$, $\alpha_2$, and $\alpha_3$. In Figure 7.1a, we have $(\alpha_1, \alpha_2, \alpha_3) - (1, 2, 3)$; in Figure 7.1b, $(\alpha_1, \alpha_2, \alpha_3) = (1, 2, 10)$; and in Figure 7.1c, $(\alpha_1, \alpha_2, \alpha_3) = (1, 2, 100)$.

As we see, the more $|\alpha_3 - \alpha_1|$ and $|\alpha_3 - \alpha_2|$ increase, the less $n_3(t)$ changes with time. Analogous conclusions can also be deduced choosing different initial conditions. Once again, this suggests to interpret $|\alpha_j - \alpha_k|$, $k \neq j$, as a sort of *inertia* of the trader $\tau_j$: the larger these differences, the bigger the tendency of $\tau_j$ to *stay quite*, that is, to keep the number of his shares constant in time, not interacting with the other traders. We could also think of $|\alpha_j - \alpha_k|^{-1}$, $k \neq j$, as a sort of *information* reaching

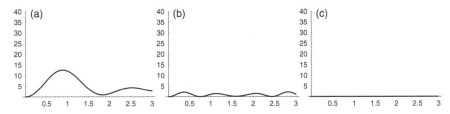

**Figure 7.1**   $n_3(t)$ for $\alpha_1 = 1$, $\alpha_2 = 2$ and $\alpha_3 = 3$ (a), $\alpha_3 = 10$ (b), and $\alpha_3 = 100$ (c).

$\tau_j$ (but not the other traders), at least if the $\alpha$s are different: if $|\alpha_j - \alpha_k|$ is large, then not much information reaches $\tau_j$, which has therefore no incoming information to optimize his interaction with the other traders. For this reason, he prefers not to buy or sell shares. This interpretation is in line with what we have deduced about the role of the parameters of the unperturbed Hamiltonians in the previous chapters, and which was analytically deduced in the simpler situation $L = 2$.

The periodic behavior of the various $n_j(t)$, which is clear for $L = 2$, is hidden for $L$ larger than two probably because of the small time interval considered in our figures. However, such a periodic or quasiperiodic behavior is ensured by the general analytical expression for the solution $a(t)$ in Equation 7.6.

As for the $L = 2$ situation, we recover that if $n_1 = n_2 = n_3 = n$, then $n_1(t) = n_2(t) = n_3(t) = n$, for all $t$. Moreover, if $n_1 \simeq n_2 \simeq n_3 \simeq n$, then all the $n_j(t)$ have small oscillations around $n$. But, if $n_1 \simeq n_2 \neq n_3$, and if $p_{ij} = 1$ for all $i, j$ with $i \neq j$, then all the functions $n_j(t)$ change *considerably* with time. The reason is that as $n_3$ differs from $n_1$ and $n_2$, it is natural to expect that $n_3(t)$ changes with time. But, as $N(t) = n_1(t) + n_2(t) + n_3(t)$ must be constant, both $n_2(t)$ and $n_1(t)$ must change in time as well. The same conclusion can be deduced also when $p_{23} = 0$ while all the other $p_{ij}$ values are equal to 1: even if $\tau_2$ does not interact directly with $\tau_3$, the fact that $\tau_1$ interacts with both $\tau_2$ and $\tau_3$, together with the fact that $N(t)$ must stay constant, implies once more that all the functions $n_j(t)$ change with time. Finally, it is clear that if $p_{13} = p_{23} = 0$, then $\tau_3$ interacts neither with $\tau_1$ nor with $\tau_2$ and, indeed, we find that $n_3(t)$ does not change with time. This is a consequence of the fact that, in this case, $[H, \hat{n}_3] = 0$, which can be seen as a *restricted version* of **R1**.

Analogous conclusions can also be deduced for five (or more) traders. In particular, Figure 7.2 shows that there is no need for all the traders to interact among them to have a nontrivial time behavior. Indeed, even if $p_{15} = p_{25} = 0$, which means that $\tau_5$ may only interact directly with $\tau_3$ and $\tau_4$, we get the plots in Figure 7.2 for $(\alpha_1, \alpha_2, \alpha_3, \alpha_4, \alpha_5) = (1, 2, 3, 4, 5)$ and $(n_1, n_2, n_3, n_4, n_5) = (40, 0, 0, 0, 0)$. We see that the number of shares of all the traders changes in time significantly.

We conclude that the interpretation of the differences of $\alpha_j$ as a sort of inertia is confirmed also by the analysis of this larger number of traders. The parameters of $H_0$ play, even for these larger systems, an interesting role in their time evolution, and must be considered if we are willing to introduce some *asymmetry* between the traders. For this reason, our results here are, somehow, more complete than those in the first part of Chapter 3, where $H_0$ was taken to be zero. The consequence of this choice was that Alice and Bob appeared to be completely equivalent, from a dynamical

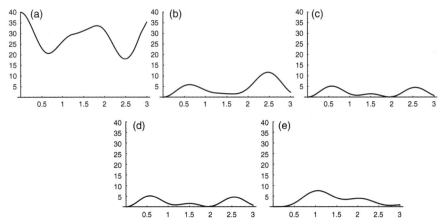

**Figure 7.2**    (a) $n_1(t)$, (b) $n_2(t)$, (c) $n_3(t)$, (d) $n_4(t)$, and (e) $n_5(t)$ for $\alpha_j$ and $n_j$ as given above.

point of view, at least when taking $M = 1$ in Equation 3.5. In order to introduce some difference between the two lovers, we could consider a Hamiltonian such as

$$H = \omega_1 \, a_1^\dagger \, a_1 + \omega_2 \, a_2^\dagger \, a_2 + \lambda \left( a_1 a_2^\dagger + a_2 \, a_1^\dagger \right),$$

which extends that in Equation 3.5. Of course, here $M = 1$. However, different behaviors of Alice and Bob can be deduced by making different choices of the frequencies $\omega_1$ and $\omega_2$.

# CHAPTER 8

# ALL-IN-ONE MODELS

In the previous chapter, we have proposed an oversimplified model of a stock market, where the only relevant ingredient of the system, and therefore of the Hamiltonian, was the number of shares of the different traders. The money was not so important (all the traders possess a huge quantity of cash), and the price of the shares was fixed, being an integral of motion. In this chapter, we consider few slightly more general (and, in our opinion, probably more interesting) models, which, however, should still be considered as models of an SSM: we are still far away from a real market but, finally, many *actors* appear in the game! The dynamical aspects of these models are discussed adopting different strategies and different approximations.

## 8.1 THE GENESIS OF THE MODEL

The basic assumptions used in this chapter to construct the SSM that we are interested in, which extend those considered in Chapter 7, are the following:

1. Our market consists of $L$ traders exchanging a single kind of share.

*Quantum Dynamics for Classical Systems: With Applications of the Number Operator*,
First Edition. Fabio Bagarello.
© 2013 John Wiley & Sons, Inc. Published 2013 by John Wiley & Sons, Inc.

2. The total number of shares, $N$, and the total amount of cash, $K$, are fixed in time.

3. A trader can interact with only a single other trader: that is, each trader feels only a *two-body interaction*.

4. The traders can buy or sell only one share (or one block of shares) in any single transaction.

5. The price of the share (or of the block of shares) changes discontinuously, with discrete steps, multiples of a given monetary unit. Again, we do not make any difference between the bid and the ask prices.

6. When the overall tendency of the market to sell a share, that is, the *market supply*, increases, the price of the share decreases, and vice versa.

7. For our convenience, the market supply is expressed in terms of natural numbers.

8. To simplify the notation, we fix the monetary unit to be equal to one.

Let us briefly comment on these conditions and their meaning, starting with Assumptions 1 and 4, which may appear rather strong. In Chapter 9, we see how to relax the first one, paying the price of some extra technical difficulties, although, concerning Assumption 4, this does not prevent at all the possibility of two traders to buy or sell more than one share (or block of shares). The only point is that the two traders must interact more than once. The reason we assume this mechanism here is that, otherwise, we would get a strongly nonlinear model, with all the extra numerical and analytical difficulties that this would imply. In fact, allowing such a possibility is similar to fixing $M$ as greater than one in Chapter 3. This will be clear later on. Assumption 2 is the first consequence of the fact that the market is closed. The existence of these two conserved quantities is used, together with **R1–R3** of Chapter 6, as a guideline to build up the Hamiltonian of the system. In other words, a closed market is, by definition, a market when the cash and the shares are not created or destroyed. Hence the two related operators, see $\hat{K}$ and $\hat{N}$ in Equation 8.6, must commute with the Hamiltonian. Going on with our analysis, Assumption 3, as we have already discussed in the previous chapter, is quite common in quantum many-body systems and simply means that a three-trader interaction is a kind of second-order (perturbative) effect, whereas a two-trader interaction can be considered as the first term in such a perturbative expansion. Or, to put it simply, it is more likely having two rather than three traders simultaneously interacting. Assumption 5 simply confirms the discrete nature of the model, as we have already discussed in Chapters 1 and 7, and introduces a single price for the share rather than two different prices for the *bidder* and the *asker*. Assumption 8 is used just to simplify the notation. Assumptions 6 and 7 provide a very simple mechanism to fix the value of the shares in terms of a global quantity, the market supply, which is a measure of the will of the various traders of the market to buy or sell the shares. Of course, there is no problem in assuming that the supply is measured in terms of natural numbers, even because this is a very natural choice in our settings, much more than requiring that it changes continuously. It is worth mentioning that, in order to construct a more complete and predictive model, we should also propose some mechanism to deduce the time evolution of the market supply itself from basic facts of the market, such as psychological and economical effects. If such a mechanism could be constructed, then Assumption 6, together with the initial conditions of the market, will also fix the price of the shares. However, this is far from being an easy task, and it is part of our future projects.

Looking at the assumptions given earlier, we immediately notice that some standard peculiarities of what in the literature is usually called a stock market (Mantegna and Stanley, 1999; Baaquie, 2004) are missing. For instance, we have not considered here any financial derivative, which, on the other hand, are the main interests in the monograph (Baaquie, 2004), which shares with this book the use of a typical quantum tool, the path integral, in the economical context. In addition, as already mentioned, we make no difference between the *bid* and *ask* prices. To be probably more realistic, ours should be better considered as a (highly nontrivial) dynamical system, which shares some common features with a stock market rather than a *detailed* market.

The *formal* Hamiltonian of the model is the following operator:

$$
\begin{cases}
\tilde{H} &= H_0 + \tilde{H}_I, \text{ where} \\
H_0 &= \sum_{l=1}^{L} \alpha_l a_l^\dagger a_l + \sum_{l=1}^{L} \beta_l c_l^\dagger c_l + o^\dagger o + p^\dagger p \\
\tilde{H}_I &= \sum_{i,j=1}^{L} p_{ij} \left[ \left( a_i^\dagger c_i^{\hat{P}} \right) \left( a_j c_j^{\dagger \hat{P}} \right) + \left( a_j^\dagger c_j^{\hat{P}} \right) \left( a_i c_i^{\dagger \hat{P}} \right) \right] \\
&\quad + (o^\dagger p + p^\dagger o),
\end{cases}
\tag{8.1}
$$

where $\hat{P} = p^\dagger p$ and the following CCR are assumed:

$$
[a_l, a_n^\dagger] = [c_l, c_n^\dagger] = \delta_{ln} \mathbb{1}, [p, p^\dagger] = [o, o^\dagger] = \mathbb{1}.
\tag{8.2}
$$

All the other commutators are zero. We also assume, as in Chapter 7 and for the same reasons, that $p_{ii} = 0$. Here, the operators $a_l^\sharp$, $p^\sharp$, $c_l^\sharp$, and $o^\sharp$

are, respectively, the *share number*, the *price*, the *cash*, and the *supply* operators. More in detail, $(a_l, a_l^\dagger)$ and $(c_l, c_l^\dagger)$ change, respectively, the number of shares and the units of cash in the portfolio of the trader $\tau_l$. The operators $(p, p^\dagger)$ change the price of the shares, whereas $(o, o^\dagger)$ change the value of the market supply. Of course, these changes are positive or negative, depending on whether creation or annihilation operators are acting. The vector states of the market are defined in the usual way:

$$\omega_{\{n\};\{k\};O;M}(\,.\,) = <\varphi_{\{n\};\{k\};O;M}\,,\,\cdot\,\varphi_{\{n\};\{k\};O;M}>,\qquad(8.3)$$

where $\{n\} = n_1, n_2, \ldots, n_L$ and $\{k\} = k_1, k_2, \ldots, k_L$ describe the number of shares and the units of cash of each trader at $t = 0$, respectively, while $O$ and $M$ fix the initial values of the market supply and of the value of the shares. Therefore,

$$\varphi_{\{n\};\{k\};O;M} := \frac{(a_1^\dagger)^{n_1}\cdots(a_L^\dagger)^{n_L}(c_1^\dagger)^{k_1}\cdots(c_L^\dagger)^{k_L}(o^\dagger)^O(p^\dagger)^M}{\sqrt{n_1!\ldots n_L!k_1!\ldots k_L!\,O!\,M!}}\,\varphi_0.\quad(8.4)$$

Here, $\varphi_0$ is the *vacuum* of the model: $a_j\varphi_0 = c_j\varphi_0 = p\varphi_0 = o\varphi_0 = 0$, for $j = 1, 2, \ldots, L$, and $n_j$, $k_j$, $O$, and $M$ are natural numbers.

The interpretation of the Hamiltonian is the key element in our approach, and follows from the general ideas discussed in Chapter 6: first, see **R1**, $H_0$ is the free Hamiltonian of the system, which contains no interaction between the various ingredients of the market. It is clear that $H_0$ commutes with all the observables of the market, that is, with all the *number operators* arising from those annihilation and creation operators appearing in Equations 8.1 and 8.2. Concerning $\tilde{H}_I$, this has been written using an extended version of **R3** of Chapter 6: the term $o^\dagger\,p$ is responsible for the supply to go up and for a simultaneous lowering of the price of the shares. This is not very different from what happened, for instance, in Chapter 3 in the (linear) description of love affairs, but with a completely different interpretation. Moreover, because of $\left(a_i^\dagger c_i^{\hat{P}}\right)\left(a_j\,c_j^{\dagger\hat{P}}\right)$, trader $\tau_i$ increases the number of shares by one unit in his portfolio but, at the same time, his cash decreases because of $c_i^{\hat{P}}$, that is, by as many units of cash as the price operator $\hat{P}$ demands. Clearly, trader $\tau_j$ behaves in the opposite way: he loses one share because of $a_j$ but his cash increases because of $(c_j^\dagger)^{\hat{P}}$. Hence, the meaning of $\tilde{H}$ in Equation 8.1 and of $\tilde{H}_I$, in particular, is now clear: it describes an SSM where two traders may buy or sell one share in each transaction, earning or losing money in this operation, and in which the price of the shares is related to the value of the supply operator as prescribed by Assumption 6. A schematic view of this market is given in Figure 8.1, where we consider (only) two traders, exchanging shares and money on the top while, at the bottom, the mechanism that fixes the price is shown, with the upward arrows carrying the information of the value of the shares to the traders. The two downward arrows represent the *feelings* of the various traders, which, all together, *construct* the market supply.

### 8.1.1   The Effective Hamiltonian

Let us go back to the Hamiltonian in Equation 8.1, which, we recall, has been called *formal*. The reason is that it suffers from a technical problem: as $c_j$ and $c_j^\dagger$ are not self-adjoint operators, it is not obvious at the first sight how to define the operators $c^{\hat{P}}$ and $(c_j^\dagger)^{\hat{P}}$; they look like non-self-adjoint, unbounded operators raised to some power, $\hat{P}$, which, by itself, is a different unbounded (but self-adjoint, at least!) operator. Maybe, the simplest way out of this problem is by replacing $\check{H}$ with an *effective* Hamiltonian, $H$, defined as

$$
\begin{cases}
H & = H_0 + H_I, \text{ where} \\
H_0 & = \sum_{l=1}^{L} \alpha_l a_l^\dagger a_l + \sum_{l=1}^{L} \beta_l c_l^\dagger c_l + o^\dagger o + p^\dagger p \\
H_I & = \sum_{i,j=1}^{L} p_{ij} \left[ \left( a_i^\dagger c_i^M \right) \left( a_j c_j^{\dagger M} \right) \right. \\
& \quad \left. + \left( a_j^\dagger c_j^M \right) \left( a_i c_i^{\dagger M} \right) \right] + (o^\dagger p + p^\dagger o),
\end{cases}
\tag{8.5}
$$

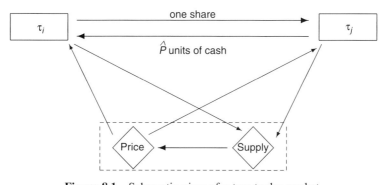

**Figure 8.1**  Schematic view of a two-trader market.

where $M$ could be taken to be some *average value* of the price operator, $<\hat{P}>$. In this section, we take $M$ as the initial value of the share,[1] and we will motivate this choice later on. This technique is quite diffused in quantum mechanics, where one replaces the *original Hamiltonian*, physically motivated but producing very difficult Heisenberg equations of motion, with an *effective Hamiltonian* which, in principle, still describes most of the original features of the system and whose related equations are easier to be solved and mathematically less *problematic*. Replacing $\hat{P}$ with $M$, however, means that we are essentially *freezing* the price of our share, removing one of the (essential) degrees of freedom of the system out of our market. Moreover, as $\hat{P}$ is also related to $\hat{O} = o^{\dagger}o$ by an integral of motion, see Equation 8.6, this also means that we are *effectively* removing a second degree of freedom from the market, keeping only the shares and the cash as the only relevant variables of the SSM. However, this is only partially true because the term $o^{\dagger} p + p^{\dagger} o$ in $H_I$ is still there and produces, as we see in a moment, a simple but not entirely trivial dynamics for $\hat{P}(t)$ and $\hat{O}(t)$. What is true is that these two operators, and $\hat{P}(t)$ in particular, are replaced by their means in that part of $H_I$ that describes the interaction between the traders: for all the traders, the price of the share is exactly $M$, and they behave as if this price does not change with time.

This approximation is relaxed later in this chapter; see Section 8.2. Having clear in mind that the Hamiltonian in Equation 8.5 gives a very rude approximation of a market, we start with this easy model as this, nonetheless, appears to be much more complete than the one considered in Chapter 7.

Three integrals of motion for our model trivially exist:

$$\hat{N} = \sum_{i=1}^{L} a_i^{\dagger} a_i, \quad \hat{K} = \sum_{i=1}^{L} c_i^{\dagger} c_i, \quad \text{and} \quad \hat{\Gamma} = o^{\dagger}o + p^{\dagger}p. \quad (8.6)$$

This can be easily checked as the CCR in Equation 8.2 implies that

$$[H, \hat{N}] = [H, \hat{\Gamma}] = [H, \hat{K}] = 0.$$

The fact that $\hat{N}$ is conserved clearly means that no new shares are introduced in the market and that no share is removed by the market. Of course, also the total amount of money must be a constant of motion as the cash is assumed to be used only to buy shares. Also because $\hat{\Gamma}$ commutes with $H$, if the mean value of $o^{\dagger}o$ increases with time, the mean value of the price operator $\hat{P} = p^{\dagger}p$ must decrease, and vice versa. This

---

[1] For this reason, to avoid unnecessary complications in Equation 8.5, we use $M$ rather than some other symbol.

is exactly the mechanism suggested in Assumption 6 at the beginning of this chapter. Moreover, also the following operators commute with $H$ and, as a consequence, are independent of time:

$$\hat{Q}_j = a_j^\dagger a_j + \frac{1}{M} c_j^\dagger c_j, \tag{8.7}$$

for $j = 1, 2, \ldots, L$. It is important to stress that the $Q_j$s are no longer integrals of motion for the original Hamiltonian, where $\hat{P}$ is not yet replaced by $M$.

The Hamiltonian (Eq. 8.5) contains a contribution, $h_{po} = o^\dagger o + p^\dagger p + (o^\dagger p + p^\dagger o)$, which is decoupled from the other terms. For this reason, it is easy to deduce the time dependence of both the price and the supply operators, and consequently of their mean values. As $h_{po}$ describes a linear dynamics, it is not hard to repeat what we did, for instance, in Chapter 3, solving the differential equations $\dot{o} = i[H, o] = i[h_{po}, o]$ and $\dot{p} = i[H, p] = i[h_{po}, p]$. These are coupled linear differential equations that can be easily solved giving

$$\begin{cases} P_r(t) = \frac{1}{2}\{M + O + (M - O)\cos(2t)\}, \\ O(t) = \frac{1}{2}\{M + O - (M - O)\cos(2t)\}. \end{cases} \tag{8.8}$$

Here, we have called $O(t) = \omega_{\{n\};\{k\};O;M}(\hat{O}(t))$ and $P_r(t) = \omega_{\{n\};\{k\};O;M}(\hat{P}(t))$, $\omega_{\{n\};\{k\};O;M}$ being the state in Equation 8.3. Notice that $P_r(0) = M$ is the initial value of the price of the shares, whereas $O$ is the value of the market supply at $t = 0$. Equation 8.8 shows that, if $O = M$, $O(t) = P_r(t) = O$ for all $t$ while if $O \simeq M$, then $O(t)$ and $P_r(t)$ are *almost* constant. In these conditions, therefore, replacing $\tilde{H}$ with $H$ does not appear to be such a drastic approximation. Moreover, the dynamical behavior arising from $H$ still looks quite interesting, as it is clear from what follows.

Our main result in Bagarello (2006) was to deduce the time evolution of the *portfolio* operator, which, for the moment, we define as

$$\hat{\Pi}_j(t) = \gamma \hat{n}_j(t) + \hat{k}_j(t). \tag{8.9}$$

Here, we have introduced the value of the share $\gamma$ *as decided by the market*, which does not necessarily coincide with the amount of money that is paid to buy the share. In a certain sense, introducing $\gamma$ is a way to give some extra freedom to the price of the shares, freedom that was partly lost when we have replaced $\hat{P}$ with $M$. As we do not need to perform such a replacement in the subsequent sections, we avoid introducing this extra parameter. Notice, however, that a different and maybe more natural definition of $\Pi_j(t)$ could be the following:

$$\Pi_j(t) = P_r(t)\hat{n}_j(t) + \hat{k}_j(t), \tag{8.10}$$

where, instead of $\gamma$, we use the time-dependent function $P_r(t)$ deduced in Equation 8.8. This is essentially our point of view in the next section.

As it is clear, $\hat{\Pi}_j(t)$ in Equation 8.9 is the sum of the total value of the shares and the cash. The fact that for each $j$ the operator $Q_j$ is an integral of motion allows us to rewrite the operator $\hat{\Pi}_j(t)$ only in terms of $\hat{n}_j(t)$ and of the initial conditions. We find

$$\hat{\Pi}_j(t) = \hat{\Pi}_j(0) + (\gamma - M)(\hat{n}_j(t) - \hat{n}_j(0)), \tag{8.11}$$

which shows that, in order to get the time behavior of the portfolio, it is enough to obtain $\hat{n}_j(t)$. We refer to Bagarello (2006) for a simple perturbation expansion of $\Pi_j(t)$ for $L = 2$, which is based on the well-known formula for operators $e^A B e^{-A} = B + [A, B] + \frac{1}{2!}[A, [A, B]] + \cdots$:

$$\hat{n}_j(t) := e^{iHt}\hat{n}_j e^{-iHt} = \hat{n}_j + it[H, \hat{n}_j] + \frac{(it)^2}{2!}\left[H, [H, \hat{n}_j]\right] + \cdots.$$

We omit this analysis here, referring to Section 8.2, when a similar expansion is considered in a more relevant context.

**The Thermodynamical Limit**    Here we concentrate on what we call the *semiclassical thermodynamical limit* of the model, that is, a nontrivial suitable limit that can be deduced for a very large number of traders, that is, when $L \to \infty$.

In this case, our model is defined by the Hamiltonian in Equation 8.5, but with $M = 1$. From an *economical* point of view, this is not a major requirement as it corresponds to a renormalization of the price of the share, which we take equal to 1: if you buy a share, then your liquidity decreases by one unit while it increases, again by one unit, if you are selling that share. Needless to say, this is strongly related to the fact that the original time-dependent price operator $\hat{P}(t)$ has been replaced by its mean value, $M$. However, fixing $M = 1$ has severe consequences in the possibility of solving the equations of motion as these become linear, and therefore, much easier to be solved, exactly because of this choice, while they are nonlinear if $M$ is taken as greater than 1. We have already found this difference in Chapter 3, and we have seen how big the difference is between a linear and nonlinear equation, especially in our scheme in which the unknowns are operator-valued.

It is clear that fixing $M = 1$ does not modify the integrals of motion found before: $\hat{N}$, $\hat{K}$ and $Q_j = \hat{n}_j + \hat{k}_j$, $j = 1, 2, \ldots, L$, as well as $\hat{\Gamma} = \hat{O} + \hat{P}$. They all commute with $H$, which we now write as

$$
\begin{cases}
H & = h + h_{po}, \text{ where} \\
h & = \sum_{l=1}^{L} \alpha_l \hat{n}_l + \sum_{l=1}^{L} \beta_l \hat{k}_l + \sum_{i,j=1}^{L} p_{ij} \left[ \left( a_i^\dagger c_i \right) \left( a_j c_j^\dagger \right) \right. \\
& \quad \left. + \left( a_j^\dagger c_j \right) \left( a_i c_i^\dagger \right) \right] \\
h_{po} & = o^\dagger o + p^\dagger p + (o^\dagger p + p^\dagger o).
\end{cases}
\tag{8.12}
$$

Incidentally, also $\hat{\Delta} := \hat{O} - \hat{P}$ commutes with $H$. For $h_{po}$, we can repeat the same arguments as before and an explicit solution can be found, which is completely independent of $h$, and coincides with that in Equation 8.8 with $M = 1$. For this reason, from now on, we identify $H$ only with $h$ in Equation 8.12 and we work only with this Hamiltonian. Let us introduce the operators

$$
X_i = a_i c_i^\dagger,
\tag{8.13}
$$

$i = 1, 2, \ldots, L$. This is (the first version of) what we call *the selling operator*: it acts on a state of the market destroying a share and creating one unit of cash in the portfolio of the trader $\tau_i$. Its adjoint $X_i^\dagger = a_i^\dagger c_i$, for obvious reasons, is called *the buying operator*. The Hamiltonian $h$ can be rewritten as

$$
h = \sum_{l=1}^{L} (\alpha_l \hat{n}_l + \beta_l \hat{k}_l) + \sum_{i,j=1}^{L} p_{ij} \left( X_i^\dagger X_j + X_j^\dagger X_i \right).
\tag{8.14}
$$

The following commutation relations can be deduced by the CCR in Equation 8.2:

$$
[X_i, X_j^\dagger] = \delta_{ij}(\hat{k}_i - \hat{n}_i), \quad [X_i, \hat{n}_j] = \delta_{ij} X_i
$$
$$
[X_i, \hat{k}_j] = -\delta_{ij} X_i,
\tag{8.15}
$$

which show how the operators $\{\{X_i, X_i^\dagger, \hat{n}_i, \hat{k}_i\}, i = 1, 2, \ldots, L\}$ are closed under commutation relations. This is quite important, as, introducing the operators $X_l^{(L)} = \sum_{i=1}^{L} p_{li} X_i$, $l = 1, 2, \ldots, L$, we get the following system of differential equations:

$$\begin{cases} \dot{X}_l = i(\beta_l - \alpha_l)X_l + 2i X_l^{(L)}(2\hat{n}_l - Q_l), \\ \dot{\hat{n}}_l = 2i \left( X_l X_l^{(L)\dagger} - X_l^{(L)} X_l^\dagger \right). \end{cases} \tag{8.16}$$

This system, as $l$ takes all the values $1, 2, \ldots, L$, is a closed system of differential equations for which a unique solution surely exists. But this is simply an existence result, not very useful in practice. More concretely, Equation 8.16 can be solved by introducing the so-called *mean-field approximation*, which essentially consists in replacing $p_{ij}$ with $\frac{\tilde{p}}{L}$, for some $\tilde{p} \geq 0$. This is, again, a standard approximation in quantum many-body systems, widely used in solid-state and in statistical mechanics; see, for instance, Thirring and Wehrl (1967) for an application in superconductivity. After this replacement, we have

$$X_l^{(L)} = \sum_{i=1}^{L} p_{li} X_i \longrightarrow \frac{\tilde{p}}{L} \sum_{i=1}^{L} X_i,$$

whose limit, for $L$ diverging, exists only in suitable topologies (Thirring and Wehrl, 1967; Bagarello and Morchio, 1992), for instance, the strong one restricted to a set of relevant states.[2] Let $\tau$ be such a topology. We define

$$X^\infty = \tau - \lim_{L \to \infty} \frac{\tilde{p}}{L} \sum_{i=1}^{L} X_i, \tag{8.17}$$

where, as it is clear, the dependence on the index $l$ is lost because of the replacement $p_{li} \to \frac{\tilde{p}}{L}$. This is a typical behavior of transactionally invariant quantum systems, where the coefficients $\{p_{li}\}$ satisfy the invariance condition $p_{li} = p_{l-i}$. The operator $X^\infty$ commutes (in some weak sense, see Thirring and Wehrl (1967)) with all the elements of $\mathcal{A}$, the *algebra of the observables* of our stock market: $[X^\infty, A] = 0$ for all $A \in \mathcal{A}$. In this limit, the system in (8.16) can be rewritten as

---

[2]The existence of this limit has a long story that goes back, for instance, to Thirring and Wehrl (1967) but that has been discussed by several authors along the years; see also Bagarello and Morchio (1992) and references therein. We are not much interested in these mathematical aspects here. We just want to mention that this limit *creates problems*, but that these problems can be solved.

$$\begin{cases} \dot{X}_l = i(\beta_l - \alpha_l)X_l + 2iX^\infty(2\hat{n}_l - Q_l), \\ \dot{\hat{n}}_l = 2i\left(X_l X^{\infty\dagger} - X^\infty X_l^\dagger\right). \end{cases} \tag{8.18}$$

This system can be analytically solved easily, at least under the hypothesis, concerning the parameters of the free Hamiltonian, that

$$\beta_l - \alpha_l =: \Phi \neq \nu, \tag{8.19}$$

for all $l = 1, 2, \ldots, L$ (but also in other and more general situations). Here, $\nu = \Phi + 4\eta - 2Q$, where

$$\eta := \tau - \lim_{L \to \infty} \frac{1}{L} \sum_{i=1}^{L} \hat{n}_i, \qquad Q := \tau - \lim_{L \to \infty} \frac{1}{L} \sum_{i=1}^{L} \hat{Q}_i.$$

We refer to Bagarello (2006) for details of this derivation and for further generalizations. Here, we just write the final result, which, calling as usual $n_l(t) = \omega_{\{n\};\{k\};O;M}(\hat{n}_l(t))$, is

$$n_l(t) = \frac{1}{\omega^2} \left\{ n_l(\Phi - \nu)^2 - 8|X_0^\infty|^2 \left( k_l(\cos(\omega t) - 1) \right.\right.$$
$$\left.\left. -n_l(\cos(\omega t) + 1) \right) \right\}. \tag{8.20}$$

We have introduced $\omega = \sqrt{(\Phi - \nu)^2 + 16|X_0^\infty|^2}$. This also allows one to find the time evolution for the portfolio of each trader, using $\Pi_l(t) = \Pi_l(0) + (\gamma - 1)(n_l(t) - n_l(0))$. Hence, at least in our assumptions, these portfolios are under control (modulo the choice of $\gamma$, which should still be fixed in some reasonable way). The easiest way would be just to fix $\gamma = M = 1$. However, this choice is not very interesting as it implies that $\Pi_l(t) = \Pi_l(0)$. Of course, the periodic behavior of $n_l(t)$ is transferred to $\Pi_l(t)$, and this gives a measure of the limits of the model presented here, as it is hard to believe that the portfolios of the traders in a real market may simply change periodically! However, in several recent applications of quantum mechanics to markets, periodic behaviors of some kind are often deduced. For instance, in Zhang and Huang (2010), a periodic behavior is deduced in connection with the Chinese market. In addition, in Schaden (2002), the author suggests that a periodic behavior is, in a sense, unavoidable anytime we deal with a closed market with few traders, whereas a nonperiodic behavior (some decay, for instance) can be obtained only in the presence of many traders. This is exactly what we have deduced in Chapters 3–5, in different contexts, where the possibility of getting such a decay is related to the presence of a large reservoir interacting with the system. Our results suggest that it is not really enough to consider a market with many traders. We also need that the majority of these traders (the reservoir) behave slightly differently than the minority, which forms the *true* market.

## 8.2  A TWO-TRADERS MODEL

As we have seen, the model analyzed in the previous section has a very strong limitation: the time evolution of the price of the share, even if formally appears in the Hamiltonian of the system, is frozen in order to get a well-defined *energy operator* (i.e., in moving from the formal operator $\tilde{H}$ to the self-adjoint Hamiltonian $H$ in Equation 8.5). Therefore, the dynamical behavior of the price operator essentially disappears in the interaction between the traders!

In this section, we correct this anomaly, and, to achieve this goal, we discuss in more detail a model based essentially on the same assumptions listed at the beginning of this chapter, but in which the market consists of only two traders, $\tau_1$ and $\tau_2$. This is to simplify the treatment, making all the technicalities much simpler.

The Hamiltonian looks very similar to the one in Equation 8.1:

$$\begin{cases} H = H_0 + H_I, \quad \text{where} \\ H_0 = \sum_{l=1}^{2} \alpha_l a_l^\dagger a_l + \sum_{l=1}^{L} \beta_l c_l^\dagger c_l + o^\dagger o + p^\dagger p \\ \tilde{H}_I = \left[ \left( a_1^\dagger c_1^{\hat{P}} \right) \left( a_2 c_2^{\dagger \hat{P}} \right) + \left( a_1 c_1^{\dagger \hat{P}} \right) \left( a_2^\dagger c_2^{\hat{P}} \right) \right] + (o^\dagger p + p^\dagger o), \end{cases}$$

$$(8.21)$$

with the standard CCR

$$[o, o^\dagger] = [p, p^\dagger] = \mathbb{1}, \qquad [a_i, a_j^\dagger] = [c_i, c_j^\dagger] = \delta_{i,j} \mathbb{1}, \qquad (8.22)$$

while all the other commutators are zero.

The states of the system are defined as in Equations 8.3 and 8.4 with $L = 2$, and the vectors $\varphi_{\{n\};\{k\};O;M}$ are eigenstates of the operators $\hat{n}_i = a_i^\dagger a_i$, $\hat{k}_i = c_i^\dagger c_i$, $i = 1, 2$, $\hat{P} = p^\dagger p$, and $\hat{O} = o^\dagger o$, respectively, with eigenvalues $n_i$, $k_i$, $i = 1, 2$, $M$, and $O$. The main improvement here is that, as we discuss in a moment, we give a rigorous meaning to the operators $c_j^P$ and $c_j^{\dagger P}$, and this allows us **not** to replace the price operator with its mean value $M$ and, as a consequence, to consider the price of the share as a real *degree of freedom* of the model, also for what concerns the interaction between $\tau_1$ and $\tau_2$; in other words, they no longer see the price of the share fixed in time. However, before defining $c_j^P$ and $c_j^{\dagger P}$, it is worth noticing that, as already stressed in Section 1.3, the observables of the market, that is, the operators $\hat{k}_i$, $\hat{n}_i$, $\hat{P}$, and $\hat{O}$, as well as some of their combinations, can be measured simultaneously. This is because these observables, which are the only relevant variables for us, do commute and, as a consequence, they admit a common set of eigenstates. Hence, no uncertainty effect arises as a consequence of our operator approach.

### 8.2.1  An Interlude: the Definition of $c^{\hat{P}}$

To make the situation simpler, we just consider here the problem in its simplest, nontrivial form: we consider two sets of independent bosonic operators $c$ and $p$, with $[c, c^\dagger] = [p, p^\dagger] = \mathbb{1}$ and $[c, p] = [c, p^\dagger] = 0$, and with a common vacuum vector, $\varphi_{0,0}$: $c\,\varphi_{0,0} = p\,\varphi_{0,0} = 0$. As usual we call

$$\varphi_{k,m} = \frac{1}{\sqrt{k!\,m!}}\,(c^\dagger)^k\,(p^\dagger)^m\,\varphi_{0,0}, \qquad (8.23)$$

where $k, m \geq 0$. We know that $\varphi_{k,m}$ is an eigenstate of $\hat{k} = c^\dagger c$ and $\hat{P} = p^\dagger p$; $\hat{k}\,\varphi_{k,m} = k\,\varphi_{k,m}$ and $\hat{P}\,\varphi_{k,m} = m\,\varphi_{k,m}$. Recalling that $c$ is a lowering operator, if $k$ is large enough, we deduce that $c\,\varphi_{k,m} = \sqrt{k}\,\varphi_{k-1,m}$, $c^2\varphi_{k,m} = \sqrt{k(k-1)}\,\varphi_{k-2,m}$, and so on. Then, it seems natural to define

$$c^{\hat{P}}\,\varphi_{k,m} := \begin{cases} \varphi_{k,m}, & \text{if } m = 0,\ \forall k \geq 0; \\ 0, & \text{if } m > k,\ \forall k \geq 0; \\ \sqrt{k(k-1)\cdots(k-m+1)}\,\varphi_{k-m,m}, & \text{if } k \geq m > 0. \end{cases}$$

$$(8.24)$$

Analogously, as $c^\dagger\varphi_{k,m} = \sqrt{k+1}\,\varphi_{k+1,m}$, $(c^\dagger)^2\varphi_{k,m} = \sqrt{(k+1)(k+2)}\,\varphi_{k+2,m}$, and so on, for all $k$ and for all $m \geq 0$, we put

$$c^{\dagger\,\hat{P}}\,\varphi_{k,m} := \begin{cases} \varphi_{k,m}, & \text{if } m=0,\ \forall k\geq 0; \\ \sqrt{(k+1)(k+2)\cdots(k+m)}\,\varphi_{k+m,m}, & \text{if } m>0. \end{cases}$$

$$(8.25)$$

Notice that $m$ appears in the right-hand side of these definitions because of the presence of the operator $\hat{P}$ in the left-hand side, and because of the eigenvalue equation $\hat{P}\varphi_{k,m} = m\,\varphi_{k,m}$. These definitions, other than natural, have two nice consequences: (i) they really essentially define the operators $c^{\hat{P}}$ and $c^{\dagger\,\hat{P}}$, as they are now defined on the vectors of an o.n. basis in the Hilbert space $\mathcal{H}$ of the system, which is the closure of the linear span of the set $\{\varphi_{k,m}, k, m \geq 0\}$.[3] Hence, there is no need to replace the Hamiltonian (Eq. 8.1) with the effective Hamiltonian (Eq. 8.5). (ii) we get an extra bonus, which suggests that Equations 8.24 and 8.25 are *good* definitions; indeed, we find that

$$(c^{\dagger})^{\hat{P}} = (c^{\hat{P}})^{\dagger},$$

as we could check verifying that $\forall\, k, m, l, s \geq 0,$

$$\langle (c^{\dagger})^{\hat{P}}\varphi_{k,m}, \varphi_{l,s}\rangle = \langle \varphi_{k,m}, c^{\hat{P}}\varphi_{l,s}\rangle.[4]$$

More relevant for us is to deduce some commutation rules that involve the operators $c^{\hat{P}}$ and $c^{\dagger\,\hat{P}}$. We claim that

$$\begin{cases} \left[\hat{P}, c^{\hat{P}}\right] = \left[\hat{P}, c^{\dagger\,\hat{P}}\right] = 0, \\[2mm] \left[\hat{k}, c^{\hat{P}}\right] = \left[c^{\dagger}c, c^{\hat{P}}\right] = -\hat{P}\,c^{\hat{P}} = -c^{\hat{P}}\,\hat{P}, \\[2mm] \left[\hat{k}, c^{\dagger\,\hat{P}}\right] = \hat{P}\,c^{\dagger\,\hat{P}} = c^{\dagger\,\hat{P}}\,\hat{P}. \end{cases} \qquad (8.26)$$

Again, we omit the proof of these rules here as they can be easily deduced applying both sides of each line above to a generic vector $\varphi_{k,m}$ of our o.n. basis and using Equations 8.24 and 8.25. We simply remark that, for instance, $[\hat{k}, c^{\hat{P}}] = -\hat{P}\,c^{\hat{P}}$ can be considered as a generalized version of $[\hat{k}, c^{l}] = -l\,c^{l}$, while $[\hat{k}, c^{\dagger\,\hat{P}}] = \hat{P}\,c^{\dagger\,\hat{P}}$ extends the commutator $[\hat{k}, c^{\dagger\,l}] = l\,c^{\dagger\,l}$.

## 8.2.2 Back to the Model

These same arguments can now be used to define the operators $c_{j}^{P}$ and $c_{j}^{\dagger\,P}$ via their action on o.n. basis of the Hilbert space $\mathcal{H}$ of the model whose

---

[3]As a matter of fact, if $X$ is an unbounded operator on $\mathcal{H}$, it is not enough to define it via its action on an o.n. basis of $\mathcal{H}$. Counterexamples exist. We should define $X$ on *any* possible o.n. basis of $\mathcal{H}$ or, more easily, on each vector of some suitable dense subspace of $\mathcal{H}$.

[4]To be more rigorous, this is not enough to prove our claim, owing to the unboundedness of $c^{\hat{P}}$. However, it is already a very good indication.

generic vector is, in analogy with Equation 8.4,

$$\varphi_{n_1,n_2;\,k_1,k_2;\,O;\,M} := \frac{(a_1^\dagger)^{n_1}(a_2^\dagger)^{n_2}(c_1^\dagger)^{k_1}(c_2^\dagger)^{k_2}(o^\dagger)^{O}(p^\dagger)^{M}}{\sqrt{n_1!\,n_2!\,k_1!\,k_2!\,O!\,M!}}\,\varphi_0. \qquad (8.27)$$

Here $n_j$, $k_j$, $O$, and $M$ are nonnegative integers, $\varphi_0$ is the *vacuum* or the *ground state* of the model, $a_j\varphi_0 = c_j\varphi_0 = p\varphi_0 = o\varphi_0 = 0$, for $j = 1, 2$, and $\mathcal{H}$ is the closure of the linear span of all these vectors. Then we have, for instance, $a_1\varphi_{n_1,n_2;\,k_1,k_2;\,O;\,M} = \sqrt{n_1}\,\varphi_{n_1-1,n_2;\,k_1,k_2;\,O;\,M}$ if $n_1 > 0$ and $a_1\varphi_{n_1,n_2;\,k_1,k_2;\,O;\,M} = 0$ if $n_1 = 0$, $a_1^\dagger\varphi_{n_1,n_2;\,k_1,k_2;\,O;\,M} = \sqrt{n_1 + 1}\,\varphi_{n_1+1,n_2;\,k_1,k_2;\,O;\,M}$, and so on. Moreover, adapting Equations 8.24 and 8.25 to the present settings, we have

$$c_1^{\hat{P}}\,\varphi_{n_1,n_2;\,k_1,k_2;\,O;\,M}$$

$$:= \begin{cases} \varphi_{n_1,n_2;\,k_1,k_2;\,O;\,M}, & \text{if } M = 0,\ \forall k_1 \geq 0; \\ 0, & \text{if } M > k_1,\ \forall k_1 \geq 0; \\ \sqrt{k_1^{\{-M\}}}\,\varphi_{n_1,n_2;\,k_1-M,k_2;\,O;\,M}, & \text{if } k_1 \geq M > 0, \end{cases} \qquad (8.28)$$

and

$$c_1^{\dagger\,\hat{P}}\,\varphi_{n_1,n_2;\,k_1,k_2;\,O;\,M}$$

$$:= \begin{cases} \varphi_{n_1,n_2;\,k_1,k_2;\,O;\,M}, & \text{if } M = 0,\ \forall k_1 \geq 0; \\ \sqrt{k_1^{\{+M\}}}\,\varphi_{n_1,n_2;\,k_1+M,k_2;\,O;\,M}, & \text{if } M > 0, \end{cases} \qquad (8.29)$$

where we have defined

$$\begin{cases} k_1^{\{-M\}} := k_1(k_1 - 1)\cdots(k_1 - M + 1), \\ k_1^{\{+M\}} := (k_1 + 1)(k_1 + 2)\cdots(k_1 + M). \end{cases} \qquad (8.30)$$

Analogous formulas hold for $c_2^{\hat{P}}$ and $c_2^{\dagger\,\hat{P}}$. These definitions have a clear *economical* interpretation: acting with $c_1^{\hat{P}}$ on $\varphi_{n_1,n_2;\,k_1,k_2;\,O;\,M}$ returns $\varphi_{n_1,n_2;\,k_1,k_2;\,O;\,M}$ itself when $M = 0$ as, in this case, the action of $c_1^{\hat{P}}$ coincides with that of the identity operator: the price of the share is zero so we do not need to pay for it. Therefore, our cash does not change, and, in fact, $k_1[new]$ (the value of $k_1$ related to the vector $c_1^{\hat{P}}\,\varphi_{n_1,n_2;\,k_1,k_2;\,O;\,M})^5$ is equal to $k_1[old]$ (the value of $k_1$ before the action of

---

[5]That is, the eigenvalue of $\hat{K}_1 = c_1^\dagger c_1$ associated with the eigenvector $c_1^{\hat{P}}\,\varphi_{n_1,n_2;\,k_1,k_2;\,O;\,M}$.

$c_1^{\hat{P}}$ on $\varphi_{n_1,n_2;\,k_1,k_2;\,O;\,M}$). If $M > k_1$, $c_1^{\hat{P}}$ destroys more *quanta of money* than $\tau_1$ really possesses: the trader $\tau_1$ is trying to set up a transaction that cannot take place! Therefore, the result of its action on the vector $\varphi_{n_1,n_2;\,k_1,k_2;\,O;\,M}$ is zero. A similar problem does not occur when we consider the action of $c_1^{\dagger\,\hat{P}}$ on $\varphi_{n_1,n_2;\,k_1,k_2;\,O;\,M}$ as, in this case, the cash in the portfolio of $\tau_1$ is created, rather than destroyed!

In the rest of the section, however, these formulas are significantly simplified by assuming that, as it is reasonable, during the transactions between $\tau_1$ and $\tau_2$, the price of the share never reaches the zero value and, moreover, that no trader tries to buy a share if he has not enough money to pay for it. Therefore, we simply rewrite Equations 8.28 and 8.29 as

$$
\begin{cases}
c_1^{\hat{P}}\,\varphi_{n_1,n_2;\,k_1,k_2;\,O;\,M} = \sqrt{k_1^{\{-M\}}}\,\varphi_{n_1,n_2;\,k_1-M,k_2;\,O;\,M}, \\[2mm]
c_1^{\dagger\,\hat{P}}\,\varphi_{n_1,n_2;\,k_1,k_2;\,O;\,M} = \sqrt{k_1^{\{+M\}}}\,\varphi_{n_1,n_2;\,k_1+M,k_2;\,O;\,M},
\end{cases}
\tag{8.31}
$$

keeping in mind, however, that if $M = 0$ or if $M > k_1$ or yet $M > k_2$, then a special care is required. The commutation rules are the standard ones; see Equation 8.2, plus the ones that extend the rules in Equation 8.26:

$$
[\hat{P}, c_j^{\hat{P}}] = [\hat{P}, c_j^{\dagger\,\hat{P}}] = 0,
\tag{8.32}
$$

and

$$
[\hat{k}_j, c_l^{\hat{P}}] = -\delta_{j,l}\,\hat{P}\,c_j^{\hat{P}}, \qquad [\hat{k}_j, c_l^{\dagger\,\hat{P}}] = \delta_{j,l}\,\hat{P}\,c_j^{\dagger\,\hat{P}},
\tag{8.33}
$$

for $j = 1, 2$.

As our market is closed, it is not surprising that the total number of shares and the total amount of cash are preserved. This is indeed proved simply computing the commutators of the total number of shares and the total cash operators, $\hat{N} = \hat{n}_1 + \hat{n}_2$ and $\hat{K} = \hat{k}_1 + \hat{k}_2$, with the Hamiltonian $H$ in Equation 8.21. Indeed, one can check that, also in this general version with $\hat{P}$ not replaced by any integer number, $[H, \hat{K}] = [H, \hat{N}] = 0$. Moreover, we can also check that $\hat{\Gamma} := \hat{O} + \hat{P}$ commutes with the Hamiltonian. This is, as already discussed in earlier sections, the mechanism that fixes the price of the share within our simplified market: when the market supply increases, the price of the share decreases, and vice versa.

As already stressed, the essential difference between the model we are considering here and the one considered previously is that now the price operator in the interaction between the traders is not replaced by its mean value. This has the first important consequence: the operators extending $Q_j$ in Equation 8.7 for this model, $a_j^\dagger a_j + \hat{P}^{-1} c_j^\dagger c_j$, $j = 1, 2$, are no longer constants of motion, and they cannot be used to simplify the computation of the portfolios of $\tau_1$ and $\tau_2$, as we have done in Section 8.1.[6] Nevertheless, we are still able to deduce, with an easy perturbative approach, the time behavior of both portfolios, at least for small values of the time.

The first step consists in deducing the time evolution of the price of the share. This computation is completely analogous to that described in Section 8.1, and is not repeated here. Again, we can deduce that $\hat{\Delta} := \hat{O} - \hat{P}$ is another constant of motion and we find that, see Equation 8.8,

$$\begin{cases} P(t) = \omega_{n_1, n_2; k_1, k_2; O; M}(\hat{P}(t)) = \frac{1}{2}\{M + O + (M - O)\cos(2t)\}, \\ O(t) = \omega_{n_1, n_2; k_1, k_2; O; M}(\hat{O}(t)) = \frac{1}{2}\{M + O - (M - O)\cos(2t)\}. \end{cases}$$
$$(8.34)$$

In Section 8.1, we have introduced another parameter, $\gamma$, which was interpreted as the price of the share *as decided by the market*, and which does not necessarily coincide with $M$. However, there was no a priori direct relation between $\gamma$ and $M$. Here we prefer not to introduce such an extra parameter and to consider directly $P(t)$ instead of its mean value $M$. In fact, we modify Equation 8.9 by defining the portfolio of the trader $\tau_j$ in a more natural way, as

$$\hat{\Pi}_j(t) = \hat{P}(t)\,\hat{n}_j(t) + \hat{k}_j(t), \qquad (8.35)$$

with $j = 1, 2$, which is just the sum of the total value of the shares and the cash of $\tau_j$. Notice that, rather than the classical function $P_r(t)$ used in Equation 8.10, here the operator $\hat{P}(t)$ appears in the definition of the portfolio.

Of course, owing to the fact that $\hat{P}(t)$ can be easily deduced solving the Heisenberg equations of motion for $p(t)$ and $p^\dagger(t)$, $\hat{\Pi}_1(t)$ is fixed by $\hat{n}_1(t)$ and $\hat{k}_1(t)$. Moreover, if we know $\hat{n}_1(t)$ and $\hat{k}_1(t)$, then we also know $\hat{n}_2(t)$ and $\hat{k}_2(t)$ as their sum must be constant, so that we can also find the analytical form of $\hat{\Pi}_2(t)$. However, this is not the only way

---

[6]We should observe that the existence of $\hat{P}^{-1}$ is guaranteed here as the eigenvalues $M$ of $\hat{P}$ are all strictly larger than zero.

to find $\hat{\Pi}_1(t)$. Another possibility follows from the fact that, as it is easy to check,

$$\dot{\hat{\Pi}}_j(t) = \hat{P}(t)\,\dot{\hat{n}}_j(t) + \dot{\hat{P}}(t)\,\hat{n}_j(t) + \dot{\hat{k}}_j(t) = \dot{\hat{P}}(t)\,\hat{n}_j(t), \qquad (8.36)$$

because $\hat{P}(t)\,\dot{\hat{n}}_j(t) + \dot{\hat{k}}_j(t) = 0$, as can be deduced by a direct computation of $\dot{\hat{n}}_j(t) = i[H, \hat{n}_j(t)]$ and $\dot{\hat{k}}_j(t) = i[H, \hat{k}_j(t)]$. Equation 8.36 shows again that, even without any need of using $Q_j$ as we did in the previous section, it is enough to know $\hat{n}_j(t)$ to deduce the time evolution of the portfolio of $\tau_j$. This is due to the fact that, as discussed earlier, $\hat{P}(t)$ can be computed explicitly, because of the ease of the model.

However, even for this two-trader model, it is not easy to deduce the exact expression for, say, $\hat{\Pi}_1(t)$. Nevertheless, a lot of information can be deduced, mainly for short time behavior, using different perturbative strategies. Here we just consider the most *direct* technique, that is, the following perturbative expansion, briefly introduced before,

$$\hat{\Pi}_1(t) = e^{iHt}\hat{\Pi}_1(0)e^{-iHt} = \hat{\Pi}_1(0) + it[H, \hat{\Pi}_1(0)]$$
$$+ \frac{(it)^2}{2!}[H, [H, \hat{\Pi}_1(0)]] + \cdots, \qquad (8.37)$$

while postponing to the next section a more detailed analysis of other strategies to produce $\hat{\Pi}_j(t)$. Computation of the various terms of this expansion, and of their mean values on the state $\omega_{n_1,n_2;\,k_1,k_2;\,O;\,M}(.)$, is based on the commutation rules we have seen earlier and produces, up to the second order in $t$, the following result:

$$\Pi_1(t) = \omega_{n_1,n_2;\,k_1,k_2;\,O;\,M}(\hat{\Pi}_1(t)) = \Pi_1(0) + t^2 n_1(O - M).$$

This formula shows that, for sufficiently small values of $t$, the value of $\Pi_1(t)$ increases with time if $O > M$, that is, if at $t = 0$ and in our units, the supply of the market is larger than the price of the share. It is further possible to check that the next term in the expansion above is proportional to $t^4 A_{n_1,n_2;k_1,k_2;M}$ where

$$A_{n_1,n_2;k_1,k_2;M} = n_1 k_1^{(+M)} k_2^{(-M)} - n_2 k_1^{(-M)} k_2^{(+M)}$$
$$+ n_1 n_2(k_1^{(+M)} k_2^{(-M)} - k_1^{(-M)} k_2^{(+M)}).$$

We avoid the details of this computation here as they are not very interesting for us, mainly because this is just a toy model that is more interesting for its general structure rather than for any realistic economical interpretation. However, we still want to stress that the expansion in Equation 8.37 gives, in principle, the expression of $\hat{\Pi}_1(t)$ at any desired order of approximation.

## 8.3 MANY TRADERS

In the previous section, we have learned how to define the operator $c^{\hat{P}}$ and its adjoint and we have used this definition in the analysis of a simple Hamiltonian written down considering only two traders. We devote this section to a slightly different model, where the stock market is made by $N$ different traders, with $N$ arbitrarily large.

In the analysis proposed here, we focus on a particular trader, $\tau$, which interacts with a group of other traders in a way that extends the interaction introduced in Equation 8.21. In other words, we divide the SSM, which, as before, is defined in terms of the number of single type of shares, the cash, the price of the shares, and the supply of the market, in two main *sets*: we call *system*, $S$, all the dynamical quantities that refer to the trader $\tau$: his/her *shares number operators*, $a$, $a^\dagger$, and $\hat{n} = a^\dagger a$, the *cash operators* of $\tau$, $c$, $c^\dagger$, and $\hat{k} = c^\dagger c$ as well as the *price operators* of the shares, $p$, $p^\dagger$, and $\hat{P} = p^\dagger p$. On the other hand, we associate with some fictitious *reservoir*, $R$, all the other quantities, that is, first of all, the *shares number operators*, $A_k$, $A_k^\dagger$, and $\hat{N}_k = A_k^\dagger A_k$ and the *cash operators*, $C_k$, $C_k^\dagger$, and $\hat{K}_k = C_k^\dagger C_k$ of the other traders. Here, $k \in \Lambda$ and $\Lambda$ is a subset of $\mathbb{N}$, which labels the traders of the market (other than $\tau$). It is clear that the cardinality of $\Lambda$ is $N - 1$. Moreover, we associate the supply of the market also with the reservoir, which is described by the operators $o_k$, $o_k^\dagger$, and $\hat{O}_k = o_k^\dagger o_k$, $k \in \Lambda$. It is clear that the reservoir we are considering here is of a different kind from those considered in Part I of the book and in Chapter 6. We keep this name as this is the standard notation adopted in the literature on the stochastic limit approach (Accardi et al., 2002), which we are going to use soon to produce the first approximation of the model. The SSM is given by the union of $S$ and $R$, and the Hamiltonian, which extends the one in Equation 8.21, is assumed here to be

$$\begin{cases} H &= H_0 + \lambda H_I, \text{ where} \\ H_0 &= \omega_a \hat{n} + \omega_c \hat{k} + \omega_p \hat{P} + \sum_{k \in \Lambda} \\ &\quad \left( \Omega_A(k) \hat{N}_k + \Omega_C(k) \hat{K}_k + \Omega_O(k) \hat{O}_k \right), \\ H_I &= \left( z^\dagger Z(f) + z Z^\dagger(\overline{f}) \right) + \left( p^\dagger o(g) + p o^\dagger(\overline{g}) \right). \end{cases} \quad (8.38)$$

Here, $\omega_a$, $\omega_c$, and $\omega_p$ are positive real numbers and $\Omega_A(k)$, $\Omega_C(k)$, and $\Omega_O(k)$ are real-valued nonnegative functions, which are related to the free time evolution of the different operators of the market. We have also introduced the following *smeared fields* of the reservoir:

$$
\begin{cases}
Z(f) = \sum_{k \in \Lambda} Z_k \, f(k) = \sum_{k \in \Lambda} A_k \, C_k^{\dagger \, \hat{P}} \, f(k), \\
Z^{\dagger}(\overline{f}) = \sum_{k \in \Lambda} Z_k^{\dagger} \, \overline{f(k)} = \sum_{k \in \Lambda} A_k^{\dagger} \, C_k^{\hat{P}} \, \overline{f(k)} \\
o(g) = \sum_{k \in \Lambda} o_k \, g(k) \\
o^{\dagger}(\overline{g}) = \sum_{k \in \Lambda} o_k^{\dagger} \, \overline{g(k)},
\end{cases}
\tag{8.39}
$$

as well as the selling operators $z = a \, c^{\dagger \, \hat{P}}$, $Z_k = A_k \, C_k^{\dagger \, \hat{P}}$ and their conjugates, the buying operators, as, for instance, $A_k$ and $C_k$ appear always in this combination both in $H_I$ and in all the computations we perform in the following. This is natural because of the *economical* meaning that we have discussed before. For instance, the action of $z$ on a fixed number vector destroys a share in the portfolio of $\tau$ and, at the same time, creates as many monetary units as $\hat{P}$ prescribes! Notice that, in $H_I$, such an operator is coupled with $Z^{\dagger}(\overline{f})$, which acts exactly in the opposite way on the traders of the reservoir: one share is created in the *cumulative* portfolio of $\mathcal{R}$ while $\hat{P}$ *quanta* of money are destroyed as they are used to pay for the share. The following nontrivial CCR are assumed:

$$
[c, c^{\dagger}] = [p, p^{\dagger}] = [a, a^{\dagger}] = \mathbb{1},
$$
$$
[o_i, o_j^{\dagger}] = [A_i, A_j^{\dagger}] = [C_i, C_j^{\dagger}] = \delta_{i,j} \, \mathbb{1},
\tag{8.40}
$$

which implies

$$
[\hat{K}_k, C_q^{\hat{P}}] = -\hat{P} \, C_q^{\hat{P}} \, \delta_{k,q}, \quad [\hat{K}_k, C_q^{\dagger \, \hat{P}}] = \hat{P} \, C_q^{\dagger \, \hat{P}} \, \delta_{k,q}.
\tag{8.41}
$$

Finally, the functions $f(k)$ and $g(k)$ in Equations 8.38 and 8.39 are sufficiently regular to allow for the sums in Equation 8.39 to be well defined, as well as the quantities that are defined later; see Equation 8.47.[7]

**Remark:** Of course, as $\tau$ can be chosen arbitrarily, the asymmetry of the model is only apparent. In fact, changing $\tau$, that is, redefining $\mathcal{S}$ and $\mathcal{R}$, we will be able, in principle, to find the time evolution of the portfolio of each trader of the SSM.

---

[7] The use of smearing functions is essential in quantum field theory, where the operators are typically operator-valued distributions (Streiter and Wightman 1964).

The interpretations suggested earlier concerning $z$ and $Z(f)$ are also based on the following results: let

$$\hat{N} := \hat{n} + \sum_{k \in \Lambda} \hat{N}_k, \quad \hat{K} := \hat{k} + \sum_{k \in \Lambda} \hat{K}_k, \quad \hat{\Gamma} := \hat{P} + \sum_{k \in \Lambda} \hat{O}_k, \quad (8.42)$$

which extend the definitions in Equation 8.6. Of course, $\hat{N}$ is associated with the total number of shares in our closed market and, therefore, is called the *total number operator*. $\hat{K}$ is the total amount of money present in the market and is called the *total cash operator*. $\hat{\Gamma}$ is the sum of the price and the *total supply* operators, $\hat{O} = \sum_{k \in \Lambda} \hat{O}_k$. It may be worth recalling that the supply operators are only related to the reservoir $\mathcal{R}$, because of our initial choice on how to split $\mathcal{S}$ and $\mathcal{R}$. This is the reason there is no contribution to the operator $\hat{O}$ coming from $\mathcal{S}$. Extending what we have deduced in earlier sections, we find that, also in the present context, the operators $\hat{N}$, $\hat{K}$, and $\hat{\Gamma}$ are constants of motion. The proof of this claim is, as always, based on the commutation rules in Equations 8.40 and 8.41. Indeed, it is not hard to check that $H$ commutes with $\hat{N}$, $\hat{K}$, and $\hat{\Gamma}$. This proves that our main motivation for introducing the Hamiltonian in Equation 8.38 is correct and satisfies the **R2** of Chapter 6: with this choice, we are constructing a closed market in which the total amount of money and the total number of shares are preserved during the time evolution and in which, if the total supply increases, then the price of the share must decrease in order for $\hat{\Gamma}$ to stay constant. Of course, as already noticed, it would be interesting to relate the changes of $\hat{O}$ to other (may be external) conditions, but this is a very hard problem that will not be considered here: in this section, we just consider the simplified point of view for which $\hat{O}$ may change in time, but we do not analyze the deep reasons as to why this happens.

The next step of our analysis should be to recover the equations of motion for the portfolio of the trader $\tau$, defined in analogy with Equation 8.35 as

$$\hat{\Pi}(t) = \hat{P}(t)\,\hat{n}(t) + \hat{k}(t). \quad (8.43)$$

It is not surprising that this cannot be done exactly, so that some perturbative technique is needed. We consider in the following subsections two quite different approaches. In particular, we first describe the so-called *stochastic limit* of the system: this approximation will produce the explicit form of the generator of the reduced semigroup of the dynamics arising from the Hamiltonian (Eq. 8.38), by taking properly into account the effect of the reservoir on the system, and this will give some interesting conditions for the *stationarity* of the model, that is, for $\hat{\Pi}(t)$ to be constant in time. We see that this is possible under certain conditions on the parameters defining the model. The second approach will make use of a sort of fixed point-like (FPL) approximation, which will produce a system of differential equations for the mean value of $\hat{\Pi}(t)$, for which the solution can be explicitly found.

### 8.3.1  The Stochastic Limit of the Model

The stochastic limit of a quantum system is a perturbative strategy, widely discussed in Accardi et al. (2002), which proved to be quite useful in the analysis of several quantum mechanical systems; see Bagarello (2005) for a review of some applications of this procedure to many-body systems. We also refer to Section 2.5.4, where a very concise list of basic facts and rules concerning the SLA, which are used now, has been given.

Here, we adopt this procedure *pragmatically*, that is, without discussing any detail of the deep theory on which the SLA is based. Again, we refer to Accardi et al. (2002) for all the details.

The first step consists in obtaining the free time evolution of the interaction Hamiltonian, which we call $H_I^0(t)$, as in the interaction representation discussed in Section 2.3.3. Using the commutation rules (Eqs 8.40 and 8.41), we find that

$$H_I^0(t) := e^{iH_0 t} H_I e^{-iHt} = z^\dagger Z(f\, e^{it\hat{\varepsilon}z}) + z\, Z^\dagger(\overline{f}\, e^{-it\hat{\varepsilon}z}) + p^\dagger o(g\, e^{it\varepsilon_0})$$
$$+ p\, o^\dagger(\overline{g}\, e^{-it\varepsilon_0}), \tag{8.44}$$

where we have defined

$$\begin{cases} \hat{\varepsilon}_Z(k) := \hat{P}(\Omega_C(k) - \omega_c) - (\Omega_A(k) - \omega_a), \\ \varepsilon_O(k) := \omega_p - \Omega_O(k), \end{cases} \tag{8.45}$$

and, for instance, $Z(f\, e^{it\hat{\varepsilon}z}) = \sum_{k \in \Lambda} f(k)\, e^{it\hat{\varepsilon}z(k)}\, Z_k$.

The next step consists of computing first $\omega\left(H_I^0\left(\frac{t_1}{\lambda^2}\right) H_I^0\left(\frac{t_2}{\lambda^2}\right)\right)$, then

$$I_\lambda(t) = \left(-\frac{i}{\lambda}\right)^2 \int_0^t dt_1 \int_0^{t_1} dt_2\, \omega\left(H_I^0\left(\frac{t_1}{\lambda^2}\right) H_I^0\left(\frac{t_2}{\lambda^2}\right)\right),$$

and finally the limit of $I_\lambda(t)$ for $\lambda \to 0$. Here, $\omega$ is the state of the market, which we take as a product state $\omega = \omega_{sys} \otimes \omega_{res}$ with $\omega_{sys}$ a Gaussian state. This means that it satisfies the following equalities:

$$\omega_{sys}(a^\sharp) = \omega_{sys}(c^\sharp) = \omega_{sys}(p^\sharp) = 0,$$

as well as

$$\omega_{sys}(a\,a) = \omega_{sys}(c\,c) = \omega_{sys}(a^\dagger a^\dagger) = \omega_{sys}(p\,p) = \omega_{sys}(p^\dagger p^\dagger) = 0.$$

In these formulas $a^\sharp$ can be $a$ or $a^\dagger$, and the same notation is adopted for $c^\sharp$ and $p^\sharp$. These conditions are obviously satisfied if $\omega_{sys}$ is a vector state analogous to the one in Equation 8.3. We do not give here the details of the computation, which is rather straightforward, but only the final result, which is obtained under the assumptions that the two functions $\varepsilon_Z(k) := \omega(\hat{\varepsilon}_Z(k))$ and $\varepsilon_O(k)$ are not identically zero. Moreover, to simplify the treatment, it is also convenient to assume that

$$\varepsilon_Y(k) = \varepsilon_Y(q) \iff k = q, \tag{8.46}$$

where $Y = Z, O$. Then, if we define the following complex constants,

$$\begin{cases} \Gamma_Z^{(a)} = \sum_{k\in\Lambda} |f(k)|^2 \,\omega_{res}(Z_k\, Z_k^\dagger) \int_{-\infty}^0 d\tau\, e^{-i\tau\varepsilon_Z(k)} \\ \Gamma_Z^{(b)} = \sum_{k\in\Lambda} |f(k)|^2 \,\omega_{res}(Z_k^\dagger\, Z_k) \int_{-\infty}^0 d\tau\, e^{i\tau\varepsilon_Z(k)} \\ \Gamma_O^{(a)} = \sum_{k\in\Lambda} |g(k)|^2 \,\omega_{res}(o_k\, o_k^\dagger) \int_{-\infty}^0 d\tau\, e^{-i\tau\varepsilon_O(k)} \\ \Gamma_O^{(b)} = \sum_{k\in\Lambda} |g(k)|^2 \,\omega_{res}(o_k^\dagger\, o_k) \int_{-\infty}^0 d\tau\, e^{i\tau\varepsilon_O(k)}, \end{cases} \tag{8.47}$$

which surely exist if $f(k)$ and $g(k)$ are regular enough, we get

$$I(t) = -t\left\{\omega_{sys}(z^\dagger z)\Gamma_Z^{(a)} + \omega_{sys}(z\,z^\dagger)\Gamma_Z^{(b)}\right.$$
$$\left. + \omega_{sys}(p^\dagger p)\Gamma_O^{(a)} + \omega_{sys}(p\,p^\dagger)\Gamma_O^{(b)}\right\}.$$

The main idea of the SLA is that this same result can be deduced, in the sense of correlators, see Section 2.5.4, using a different self-adjoint, time-dependent operator $H^{(ls)}(t)$, the so-called *stochastic limit* Hamiltonian.

Let us take

$$H^{(ls)}(t) = z^\dagger\left(Z^{(a)}(t) + Z^{(b)\dagger}(t)\right) + z\left(Z^{(a)\dagger}(t) + Z^{(b)}(t)\right)$$
$$+ p^\dagger\left(o^{(a)}(t) + o^{(b)\dagger}(t)\right) + p\left(o^{(a)\dagger}(t) + o^{(b)}(t)\right), \tag{8.48}$$

where the new operators of the *limiting reservoir* introduced here are assumed to satisfy the following commutation rules:

$$\left[Z^{(a)}(t), Z^{(a)\dagger}(t')\right] = \Gamma_Z^{(a)} \delta_+(t - t'),$$

$$\left[Z^{(b)}(t), Z^{(b)\dagger}(t')\right] = \Gamma_Z^{(b)} \delta_+(t - t'),$$

(8.49)

and

$$\left[o^{(a)}(t), o^{(a)\dagger}(t')\right] = \Gamma_O^{(a)} \delta_+(t - t'),$$

$$\left[o^{(b)}(t), o^{(b)\dagger}(t')\right] = \Gamma_O^{(b)} \delta_+(t - t'),$$

(8.50)

if $t \geq t'$. The time ordering is crucial, and $\delta_+$ is essentially the Dirac delta function except for a normalization that arises because of the time ordering we consider here (Accardi et al., 2002). The only property of $\delta_+$ that we need is the following: $\int_0^t \delta_+(t - \tau) h(\tau) d\tau = h(t)$, for all test functions $h(t)$.

Now, let $\Psi_0$ be the vacuum of the operators $Z^{(a)}(t)$, $Z^{(b)}(t)$, $o^{(a)}(t)$, and $o^{(b)}(t)$. This means that $Z^{(a)}(t)\Psi_0 = Z^{(b)}(t)\Psi_0 = o^{(a)}(t)\Psi_0 = o^{(b)}(t)\Psi_0 = 0$ for all $t \geq 0$. It is clear that, in general, $\Psi_0$ has nothing to do with $\omega_{res}$, the original state of the reservoir. In Accardi et al. (2002), it is discussed that the Hilbert space of the reservoir must also be replaced during this procedure. Then, if we consider a new state, $\Omega(.) = \omega_{sys}(.) \otimes < \Psi_0, . \Psi_0 >$, and we compute

$$J(t) = (-i)^2 \int_0^t dt_1 \int_0^{t_1} dt_2 \, \Omega\left(H^{(ls)}(t_1) H^{(ls)}(t_2)\right),$$

using Equations 8.49 and 8.50, we conclude that $J(t) = I(t)$. In Accardi et al. (2002), this is understood as an indication that, at a first relevant order, $H^{(ls)}(t)$ can be used to get the wave operator $U_t$ that describes the time evolution of the system. We use $H^{(ls)}(t)$ to construct the wave operator as $U_t = \mathbb{1} - i \int_0^t H^{(ls)}(t') U_t'$, and then to deduce the following commutation rules:

$$\left[Z^{(a)}(t), U_t\right] = -i\Gamma_Z^{(a)} z \, U_t, \quad \left[Z^{(b)}(t), U_t\right] = -i\Gamma_Z^{(b)} z^\dagger \, U_t, \quad (8.51)$$

and

$$\left[o^{(a)}(t), U_t\right] = -i\Gamma_O^{(a)} p \, U_t, \quad \left[o^{(b)}(t), U_t\right] = -i\Gamma_O^{(b)} p^\dagger \, U_t, \quad (8.52)$$

which are obtained by making use of the *time consecutive principle* (Accardi et al., 2002).

We are now ready to get the expression of the generator. Let $X$ be a generic observable of the system, that is, in our present context, some dynamical variable related to the trader $\tau$. Let $\mathbb{1}_r$ be the identity operator of the reservoir. Then the time evolution of $X \otimes \mathbb{1}_r$ in the interaction picture is given by $j_t(X \otimes \mathbb{1}_r) = U_t^\dagger (X \otimes \mathbb{1}_r) U_t$, so that

$$\partial_t j_t(X \otimes \mathbb{1}_r) = i U_t^\dagger [H^{(ls)}(t), X \otimes \mathbb{1}_r] U_t$$

Using now the commutators in Equations 8.51 and 8.52, and recalling that $\Psi_0$ is annihilated by all the *new* reservoir operators, we find that

$$\Omega\left(\partial_t j_t(X \otimes \mathbb{1}_r)\right) = \Omega(U_t^\dagger\{\Gamma_Z^{(a)}[z^\dagger, X] z - \overline{\Gamma_Z^{(a)}} z^\dagger [z, X]$$
$$+ \Gamma_Z^{(b)}[z, X] z^\dagger - \overline{\Gamma_Z^{(b)}} z [z^\dagger, X] +$$
$$+ \Gamma_O^{(a)}[p^\dagger, X] p - \overline{\Gamma_O^{(a)}} p^\dagger [p, X] + \Gamma_O^{(b)}[p, X] p^\dagger$$
$$- \overline{\Gamma_O^{(b)}} p [p^\dagger, X]\} U_t),$$

which, together with the equality $\Omega\left(\partial_t j_t(X \otimes \mathbb{1}_r)\right) = \Omega(j_t(L(X \otimes \mathbb{1}_r)))$, gives us the following expression of the generator:

$$L(X \otimes \mathbb{1}_r) = \Gamma_Z^{(a)}[z^\dagger, X] z - \overline{\Gamma_Z^{(a)}} z^\dagger [z, X] + \Gamma_Z^{(b)}[z, X] z^\dagger$$
$$- \overline{\Gamma_Z^{(b)}} z [z^\dagger, X] + \Gamma_O^{(a)}[p^\dagger, X] p - \overline{\Gamma_O^{(a)}} p^\dagger [p, X]$$
$$+ \Gamma_O^{(b)}[p, X] p^\dagger - \overline{\Gamma_O^{(b)}} p [p^\dagger, X] \tag{8.53}$$

Applying this formula to the operators we are interested in, we find, after few computations, that

$$L(\hat{n} \otimes \mathbb{1}_r) = 2\Re\{\Gamma_Z^{(b)}\} z z^\dagger - 2\Re\{\Gamma_Z^{(a)}\} z^\dagger z, \tag{8.54}$$

and

$$L(\hat{k} \otimes \mathbb{1}_r) = -2 \hat{P} \Re\{\Gamma_Z^{(b)}\} z z^\dagger + 2 \hat{P} \Re\{\Gamma_Z^{(a)}\} z^\dagger z, \tag{8.55}$$

which, in particular, shows that $L(\hat{k} \otimes \mathbb{1}_r) + \hat{P} L(\hat{n} \otimes \mathbb{1}_r) = 0$. This equality is similar to the one we have used while writing Equation 8.36.

Finally we find, using these results and recalling that $\hat{\Pi}(t) = \hat{P}(t)\hat{n}(t) + \hat{k}(t)$,

$$L(\hat{\Pi} \otimes \mathbb{1}_r) = 2\,(\Re\{\Gamma_O^{(b)}\} - \Re\{\Gamma_O^{(a)}\})\,\hat{P}\,\hat{n} + 2\,\Re\{\Gamma_O^{(b)}\}\,\hat{n}. \qquad (8.56)$$

The first remark is that, in the stochastic limit, even if the time dependence of $\hat{n}$ and $\hat{k}$ depends on $\Gamma_Z^{(a)}$ and $\Gamma_Z^{(b)}$, the time evolution of $\hat{\Pi}$ in a first approximation does not! In fact, Equation 8.56 shows that it only depends on $\Gamma_O^{(a)}$ and $\Gamma_O^{(b)}$.

The earlier equations show that even after the stochastic limit has been taken, it is quite difficult to produce a closed set of differential equations. Nevertheless, it is quite easy to deduce conditions for the stationarity of the market. This is exactly what we will discuss next.

We begin noticing that, for instance, we have

$$2\Re\{\Gamma_O^{(a)}\} = \sum_{k\in\Lambda} |g(k)|^2\,\omega_{\text{res}}(o_k\,o_k^\dagger) \int_{\mathbb{R}} d\tau\, e^{-i\tau\,\varepsilon_O(k)}$$

$$= 2\pi \sum_{k\in\Lambda} |g(k)|^2\,\omega_{\text{res}}(o_k\,o_k^\dagger)\,\delta(\varepsilon_O(k)) \qquad (8.57)$$

and analogously we find that

$$\Re\{\Gamma_O^{(b)}\} = \pi \sum_{k\in\Lambda} |g(k)|^2\,\omega_{\text{res}}(o_k^\dagger\,o_k)\,\delta(\varepsilon_O(k)),$$

while

$$\Re\{\Gamma_Z^{(a)}\} = \pi \sum_{k\in\Lambda} |f(k)|^2\,\omega_{\text{res}}(Z_k\,Z_k^\dagger)\,\delta(\varepsilon_Z(k))$$

and

$$\Re\{\Gamma_Z^{(b)}\} = \pi \sum_{k\in\Lambda} |f(k)|^2\,\omega_{\text{res}}(Z_k^\dagger\,Z_k)\,\delta(\varepsilon_Z(k)).$$

Therefore, as $[o_k, o_k^\dagger] = \mathbb{1}$ and, as a consequence, $\omega_{\text{res}}(o_k\,o_k^\dagger) - \omega_{\text{res}}(o_k^\dagger\,o_k) = \omega_{\text{res}}(\mathbb{1}) = 1$, we find that $\Re\{\Gamma_O^{(b)}\} - \Re\{\Gamma_O^{(a)}\} = -\pi \sum_{k\in\Lambda} |g(k)|^2\delta(\varepsilon_O(k))$. The conclusion now follows from Equation 8.56: the portfolio of $\tau$ is *stationary* (in our approximation) when the function $\varepsilon_O(k)$ has no zero for $k \in \Lambda$. Indeed, if this is the case, we deduce that $L(\hat{\Pi} \otimes \mathbb{1}_r) = 0$. As $\varepsilon_O(k) = \omega_p - \Omega_O(k)$, this means that if the free dynamics of the price and the supply are based on substantially different quantities, then the portfolio of $\tau$ keeps its original value, even if the operators $\hat{n}(t)$ and $\hat{k}(t)$ may separately change with time. This is an interesting result as it can be summarized just stating that, within the approximation we are considering here, the fact that $\hat{\Pi}(t)$ depends or not on time is only related to a given *equilibrium*, if any, between the free price Hamiltonian, $\omega_p\, p^\dagger\, p$, and the free supply Hamiltonian, $\sum_{k\in\Lambda}\Omega_O(k)\, c_k^\dagger\, c_k$: this is another evidence of the fact that the free Hamiltonian of an interacting system plays, in general, a very important role in the time evolution of the system.

A similar analysis can also be carried out to get conditions for the equilibrium of $\hat{n}(t)$ and $\hat{k}(t)$. Because of Equations 8.54 and 8.55, and because of the known time evolution of $\hat{P}(t)$, $\hat{n}(t)$ is constant if and only if $\hat{k}(t)$ is constant, and for this to be true, the function $\varepsilon_Z(k)$ must be different from zero for each $k \in \Lambda$. On the other hand, if at least one zero of $\varepsilon_Z(k)$ exists in $\Lambda$, then $\hat{n}(t)$ and $\hat{k}(t)$ are, in general, time-dependent operators.

### 8.3.2 The FPL Approximation

As we have seen, the SLA discussed in the previous section is useful mainly to deduce conditions for the stationarity of the dynamics of the interacting system. However, as we are more interested in its dynamical behavior, we consider here a completely different approach, which produces a nice approximation of the time evolution of the portfolio of the trader $\tau$.

We start reminding that, given a generic operator $X$, its time evolution in the Heisenberg representation is (formally) given by $X(t) = e^{iHt}\, X\, e^{-iHt}$, and it satisfies the following well-known Heisenberg equation of motion (Eq. 2.2): $\dot{X}(t) = ie^{iHt}[H, X]e^{-iHt} = i[H, X(t)]$. In the attempt at deducing the analytical expression for $\hat{\Pi}(t)$, the following differential equations are obtained:

$$
\begin{cases}
\dfrac{d\hat{n}(t)}{dt} = i\lambda\left(-z^\dagger(t)\,Z(f,t) + z(t)\,Z^\dagger(\overline{f},t)\right), \\[2ex]
\dfrac{d\hat{k}(t)}{dt} = i\lambda\,\hat{P}(t)\left(z^\dagger(t)\,Z(f,t) - z(t)\,Z^\dagger(\overline{f},t)\right), \\[2ex]
\dfrac{d\hat{P}(t)}{dt} = i\lambda\left(p(t)\,o^\dagger(\overline{g},t) - p^\dagger(t)\,o(g,t)\right), \\[2ex]
\dfrac{dz(t)}{dt} = i\left(\hat{P}(t)\omega_c - \omega_a\right)z(t) + i\lambda[z^\dagger(t), z(t)]\,Z(f,t), \\[2ex]
\dfrac{dZ(f,t)}{dt} = i\,Z\left((\hat{P}(t)\Omega_C - \Omega_A)f, t\right) + i\lambda\, z(t)\,[Z^\dagger(\overline{f},t), Z(f,t)].
\end{cases}
$$

$$(8.58)$$

where we have defined $Z(f,t) := e^{iHt} Z(f) e^{-iHt}$, $Z((\hat{P}(t)\Omega_C - \Omega_A)$ $f,t) = \sum_{k \in \Lambda} (\hat{P}(t)\Omega_C(k) - \Omega_A(k)) f(k) Z_k(t), o(g,t) = e^{iHt} o(g) e^{-iHt}$, and so on.

It is clear that Equation 8.58 is not closed as, for instance, the differential equation for $\hat{P}(t)$ involves $p(t)$, $o(g,t)$, and their adjoint. This is not a major problem because, as we have widely discussed before, it is quite easy to deduce the time evolution of the price operator $\hat{P}$ with no approximation at all. This is because $p(t)$ (and $o(g,t)$) can be found explicitly and $P(t)$ and $O(g,t)$ can be deduced in analogy with Equation 8.34. In particular, even if these operators can be found under more general conditions, we now require that the coefficients in $H$ satisfy some extra requirement, which are only useful to simplify the computations. For instance, we assume that $\Omega_O(k)$ is constant in $k$, $\Omega_O = \Omega_O(k)$ for all $k \in \Lambda$, and that $\omega_p = \sum_{k \in \Lambda} |g(k)|^2 = \lambda = \Omega_O$. Then, we get $p(t) = \frac{1}{2} \left( p(e^{-2i\lambda t} + 1) + o(g)(e^{-2i\lambda t} - 1) \right)$ and $\hat{P}(t) = p^\dagger(t) p(t)$. As $\hat{P}(t)$ depends only on the operators $p$ and $o$, but not on $a, c$, and so on, and as we are interested in the mean value of some number operators on a vector state $\omega$ generalizing the one in Equation 8.3, we replace system (8.58) with its *semiclassical approximation*

$$
\begin{cases}
\dfrac{d\hat{n}(t)}{dt} = i\lambda \left( -z^\dagger(t) Z(f,t) + z(t) Z^\dagger(\overline{f},t) \right), \\[2ex]
\dfrac{d\hat{k}(t)}{dt} = i\lambda P(t) \left( z^\dagger(t) Z(f,t) - z(t) Z^\dagger(\overline{f},t) \right), \\[2ex]
\dfrac{dz(t)}{dt} = i \left( P(t)\omega_c - \omega_a \right) z(t) + i\lambda [z^\dagger(t), z(t)] Z(f,t), \\[2ex]
\dfrac{dZ(f,t)}{dt} = i Z \left( (P(t)\Omega_C - \Omega_A) f, t \right) + i\lambda z(t) [Z^\dagger(\overline{f},t), Z(f,t)],
\end{cases}
$$

$$\text{(8.59)}$$

where

$$
P(t) = \omega(\hat{P}(t)) = \frac{1}{2} [(M + O) + (M - O) \cos(2\lambda t)]. \qquad \text{(8.60)}
$$

In order to simplify the analysis of this system even more, it is also convenient to assume that both $\Omega_C(k)$ and $\Omega_A(k)$ are constant in $k$. Indeed, under this assumption, the last two equations in (8.59) form by themselves a closed system of differential equations in the nonabelian variables $z(t)$ and $Z(f, t)$:

$$\begin{cases} \dfrac{dz(t)}{dt} = i\left(P(t)\omega_c - \omega_a\right) z(t) + i\lambda \, Z(f, t) \, [z^\dagger(t), z(t)], \\[2mm] \dfrac{dZ(f, t)}{dt} = i\left(P(t)\Omega_C - \Omega_A\right) Z(f, t) + i\lambda \, z(t) \, [Z^\dagger(\overline{f}, t), Z(f, t)]. \end{cases}$$

(8.61)

Getting the exact solution of Equation 8.59, with Equation 8.61 replacing the two last equations, is a hard job. Nevertheless, this set of equations appears to be a good starting point for finding an approximated solution of the original dynamical problem. Indeed, a natural approach consists of taking the first nontrivial contribution of the system, as usually done in perturbation theory. This means that, in Equation 8.61, the contributions containing the commutators must be neglected as they are proportional to the interaction parameter $\lambda$, assumed to be small, although $i\left(P(t)\omega_c - \omega_a\right) z(t)$ and $i\left(P(t)\Omega_C - \Omega_A\right) Z(f, t)$ which, on the other hand, do not depend on $\lambda$, give the first relevant contribution. Moreover, in order not to trivialize the system, we have to keep the first two equations (in 8.59) as they are; if we also simply put $\lambda = 0$ here, in fact, we would deduce that the time evolution of both $\hat{n}(t)$ and $\hat{k}(t)$ is trivial: $\hat{n}(t) = \hat{n}$ and $\hat{k}(t) = \hat{k}$, for all $t$. With this choice we get

$$\begin{cases} \dfrac{d\hat{n}(t)}{dt} = i\lambda\left(-z^\dagger(t)\, Z(f, t) + z(t)\, Z^\dagger(\overline{f}, t)\right), \\[2mm] \dfrac{d\hat{k}(t)}{dt} = i\lambda\, P(t)\left(z^\dagger(t)\, Z(f, t) - z(t)\, Z^\dagger(\overline{f}, t)\right), \\[2mm] \dfrac{dz(t)}{dt} = i\left(P(t)\omega_c - \omega_a\right) z(t), \\[2mm] \dfrac{dZ(f, t)}{dt} = i\left(P(t)\Omega_C - \Omega_A\right) Z(f, t). \end{cases}$$

(8.62)

However, we can check that this approximation is too strong: in fact, solving Equation 8.62, we can see that, even if the operators $\hat{n}(t)$ and $\hat{k}(t)$ have a nontrivial dynamics, their mean values, $n(t) = \omega(\hat{n}(t))$ and $k(t) = \omega(\hat{k}(t))$, turn out to stay constant in time, so that the time behavior of the (semiclassical) portfolio $\Pi(t) = P(t)\, n(t) + k(t) = P(t)\, n + k$ is uniquely given by the function $P(t)$ in Equation 8.60.

To see this, we first observe that $z(t)$ and $Z(f, t)$ in Equation 8.62 are

$$z(t) = z\, e^{i\chi(t)}, \qquad Z(f, t) = Z(f)\, e^{i\tilde{\chi}(t)}, \tag{8.63}$$

where

$$\chi(t) = \alpha t + \beta \sin(2\lambda t), \qquad \tilde{\chi}(t) = \tilde{\alpha} t + \tilde{\beta} \sin(2\lambda t) \qquad (8.64)$$

with

$$
\begin{cases}
\alpha = \frac{1}{2}((M + O)\omega_c - 2\omega_a), & \beta = \dfrac{\omega_c}{4\lambda}(M - O), \\[2mm]
\tilde{\alpha} = \dfrac{1}{2}((M + O)\Omega_C - 2\Omega_A), & \tilde{\beta} = \dfrac{\Omega_C}{4\lambda}(M - O).
\end{cases}
\qquad (8.65)
$$

Our claim is now an immediate consequence of Equation 8.63. Indeed, from the first equation in Equation 8.62, taking its mean value on the number vector state $\omega$, we find

$$\dot{n}(t) = \frac{d}{dt}\,\omega(\hat{n}(t)) = \omega\left(\frac{d}{dt}\hat{n}(t)\right) = i\lambda\left\{-\omega\left(z^\dagger(t)Z(f,t)\right)\right.$$

$$\left. +\omega\left(z(t)Z^\dagger(\overline{f},t)\right)\right\} = 0$$

as, for instance, $\omega\left(z^\dagger(t)Z(f,t)\right) = e^{-i(\chi(t)-\tilde{\chi}(t))}\omega(z^\dagger Z(f)) = 0$. Analogously, we find that $\dot{k}(t) = \frac{d}{dt}\omega(\hat{k}(t)) = 0$. Therefore, we deduce that $n(t) = n$ and $k(t) = k$, as claimed earlier.

**Remark:** In a certain sense, this result produces a link between the present approach and the SLA discussed in the previous section. Indeed, replacing Equation 8.59 with Equation 8.62, we obtain a stationary behavior for $n(t)$ and $k(t)$. An analogous behavior was deduced, using the SLA, if $\varepsilon_Z(k)$ has no zero. However, these two different approximations cannot be directly compared. The reason is the following: in Section 8.3.1, we had to require that $\varepsilon_Z(k)$ and $\varepsilon_O(k)$ were not identically zero. This is crucial to ensure the existence of $\lim_{\lambda,0} I_\lambda(t)$. In the present approximation, we are requiring that both $\Omega_C(k)$ and $\Omega_A(k)$ are constants in $k$ so that, see Equation 8.45, we would get $\varepsilon_Z(k) = P(t)(\Omega_C - \omega_c) - (\Omega_A - \omega_a)$, which may have some zero in $k$ only if it is identically zero. In other words, we are working here assuming conditions for which the SLA does not work. Vice versa, if we are under the conditions of the previous subsection, then Equation 8.59 cannot be easily solved! Hence, the two approximations cover different situations.

A better approximation can be constructed. Again the starting point is Equation 8.61, for which we now construct iteratively a solution, stopping at the second order. In other words, we take $z_0(t)$ and $Z_0(f, t)$ as in Equation 8.63, $z_0(t) = z\, e^{i\chi(t)}$ and $Z_0(f, t) = Z(f)\, e^{i\bar{\chi}(t)}$, and then we look for the next approximation by considering the following system:

$$\begin{cases} \dfrac{dz_1(t)}{dt} = i\left(P(t)\omega_c - \omega_a\right) z_0(t) + i\lambda\, Z_0(f, t)\, [z_0^\dagger(t), z_0(t)], \\ \dfrac{dZ_1(f, t)}{dt} = i\,(P(t)\Omega_C - \Omega_A)\, Z_0(f) + i\lambda\, z_0(t)\, [Z_0^\dagger(\overline{f}, t), Z_0(f, t)], \end{cases}$$

which can be still written as

$$\begin{cases} \dfrac{dz_1(t)}{dt} = i\left(P(t)\omega_c - \omega_a\right) z_0(t) + i\lambda\, Z_0(f, t)\, [z_0^\dagger, z_0], \\ \dfrac{dZ_1(f, t)}{dt} = i\,(P(t)\Omega_C - \Omega_A)\, Z_0(f) + i\lambda\, z_0(t)\, [Z_0^\dagger(\overline{f}), Z_0(f)]. \end{cases} \tag{8.66}$$

Those readers who are familiar with the proof of the existence of a fixed point for strict contractions in Functional Analysis may see that the idea behind the approximation we are constructing here is very close to that one. We get approximated solutions, $f_k$, labeled by an integer $k$, whose $k$-th approximation can be found from $f_{k-1}$ using some recurrence relation. This is the reason we have called this section *fixed point-like approximation*. Equation 8.66 can be solved and the solution can be written as

$$\left.\begin{aligned} z_1(t) &= z\,\eta_1(t) + Z(f)\,[z^\dagger, z]\,\eta_2(t), \\ Z_1(f, t) &= Z(f)\,\tilde{\eta}_1(t) + z\,[Z(\overline{f})^\dagger, Z(f)]\,\tilde{\eta}_2(t), \end{aligned}\right\} \tag{8.67}$$

where we have introduced the following functions

$$\begin{cases} \eta_1(t) = 1 + i \int_0^t (P(t')\omega_c - \omega_a)\, e^{i\chi(t')}\, dt', \; \eta_2(t) = i\lambda \int_0^t e^{i\bar{\chi}(t')}\, dt' \\ \tilde{\eta}_1(t) = 1 + i \int_0^t (P(t')\Omega_C - \Omega_A)\, e^{i\bar{\chi}(t')}\, dt', \; \tilde{\eta}_2(t) = i\lambda \int_0^t e^{i\chi(t')}\, dt' \end{cases} \tag{8.68}$$

It is not a big surprise that this approximated solution does not share with $z(t)$ and $Z(f, t)$ all their properties. In particular, while, for instance, $[z(t), Z(f, t)] = 0$ for all $t$, $z_1(t)$ and $Z_1(f, t)$ do not commute.

It is easy to conclude that the mean values of the first two equations in (8.62) can be written as

$$
\begin{cases}
\dot{n}(t) = \dfrac{dn(t)}{dt} = -2\lambda\Im\left\{\omega\left(z(t)\,Z^{\dagger}(\overline{f},t)\right)\right\}, \\[2mm]
\dot{k}(t) = \dfrac{dk(t)}{dt} = 2\lambda\,P(t)\Im\left\{\omega\left(z(t)\,Z^{\dagger}(\overline{f},t)\right)\right\},
\end{cases}
\tag{8.69}
$$

which, in particular, implies a well-known identity: $P(t)\dot{n}(t) + \dot{k}(t) = 0$ for all $t$. A consequence of this condition is that the time derivative of the portfolio operator of $\tau$ satisfies the following equation: $\dot{\Pi}(t) = \dot{P}(t)\,n(t)$. We recall that this equality was already deduced several times in this chapter, in very different ways and in similar forms. It should be remarked that, because of this relation, and recalling that, when $M = O$, $P(t) = P(0) = M$; hence, when $M = O$, the dynamics of the portfolio of $\tau$ is trivial: $\dot{\Pi}(t) = 0$ and, as a consequence, $\Pi(t) = \Pi(0)$, even if both $n(t)$ and $k(t)$ may change in time.

At this stage, we put $z_1(t)$ and $Z_1(f, t)$ in Equation 8.69. If $\omega$ is the usual number state, and if we call for simplicity

$$
\begin{cases}
\omega(1) := \omega\left(zz^{\dagger}\,[Z^{\dagger}(\overline{f}), Z(f)]\right\} \\[1mm]
\omega(2) := \omega\left(Z(f)Z^{\dagger}(\overline{f})\,[z^{\dagger}, z]\right\} \\[1mm]
r(t) = \omega(1)\,\eta_1(t)\,\overline{\tilde{\eta}_2(t)} + \omega(2)\,\eta_2(t)\,\overline{\tilde{\eta}_1(t)}
\end{cases}
\tag{8.70}
$$

then we get

$$
\begin{cases}
n(t) = n - 2\lambda\Im\left\{\int_0^t r(t')\,dt'\right\}, \\[2mm]
k(t) = k + 2\lambda\Im\left\{\int_0^t P(t')\,r(t')\,dt'\right\}.
\end{cases}
\tag{8.71}
$$

The time dependence of the portfolio can now be written as

$$
\Pi(t) = \Pi(0) + \delta\Pi(t),
\tag{8.72}
$$

where

$$
\begin{aligned}
\delta\Pi(t) = {}& n(O - M)\sin^2(\lambda t) \\[1mm]
& + \left(-2\lambda\,\Im\left\{\int_0^t r(t')\,dt'\right\}\right)\left(M + (O - M)\sin^2(\lambda t)\right) \\[1mm]
& + 2\lambda\,\Im\left\{\int_0^t P(t')\,r(t')\,dt'\right\}\right),
\end{aligned}
\tag{8.73}
$$

which gives the variation of the portfolio of $\tau$ in time. We observe that, as it is expected, $\delta\Pi(t) = 0$ if $\lambda = 0$. We would also like to stress that because of the appearance of $\sin^2(\lambda t)$, Equation 8.73 does not appear to be a simple expansion of $\Pi(t)$ in powers of $\lambda$. Because of the definitions of $r(t)$, we deduce that the solution $\delta\Pi(t)$ in Equation 8.73 is a nontrivial combination of functions having some periodicity. It is natural, therefore, to expect that $\delta\Pi(t)$ presents some quasiperiodicity, but this feature is only partly visible in the plots produced here because of the small time interval considered.

In the last part of this section, we look for particular solutions of this system under special conditions. The first remark concerning Equation 8.73 is the following: because of the analytical expression of $\delta\Pi(t)$, if $O > M$, it is more likely for $\tau$ to have a positive $\delta\Pi(t)$ if the number of the shares $n$ in his/her portfolio at time $t = 0$ is large. This is because, for $n$ large enough, the first contribution to $\delta\Pi(t)$ in Equation 8.73, $n(O - M)\sin^2(\lambda t)$, is larger than the other terms. Therefore, if at $t = 0$ the supply of the market is larger than the price of the share, then *for a trader with many shares, it is easy to become even richer*! If, on the contrary, $O < M$, having a large number of shares does not automatically produce an increment of the portfolio.

Coefficients $\omega(1)$ and $\omega(2)$ can be found explicitly and depend on the initial conditions of the market. If, for simplicity's sake, we consider here $\Lambda = \{k_o\}$, that is, if the reservoir consists of just another trader interacting with $\tau$, then we get

$$\omega(1) = |f(k_o)|^2 (1 + n) k^{\{-M\}} \left( n_o k_o^{\{+M\}} - (1 + n_o) k_o^{\{-M\}} \right),$$

and

$$\omega(2) = |f(k_o)|^2 (1 + n_o) k_o^{\{-M\}} \left( n k^{\{+M\}} - (1 + n) k^{\{-M\}} \right).$$

It is clear that these coefficients coincide if $k = k_o$ and $n = n_o$.

Let us first fix $M = 1$, $O = 2$, $\lambda = 1$, $\omega_a = \omega_c = 1$, and $\Omega_A = \Omega_C = 2$. Then the plots of $\delta\Pi(t)$ in Figure 8.2, in which $n$ is fixed to be 10, are related to the following different values of $\omega(1)$ and $\omega(2)$: $(\omega(1), \omega(2)) = (1, 1), (1, 10), (10, 1)$.

The plots do not change much if we fix $n = 5$ and, surprisingly enough, also the ranges of variations of $\delta\Pi(t)$ essentially coincide with those given earlier: $n$ does not seem to be so important in this case! In Figure 8.3, we plot $\delta\Pi(t)$ in the same conditions as before, but for $n = 5$.

From both these figures we see that, for trader $\tau$, the most convenient situation is when $(\omega(1), \omega(2)) = (1, 10)$: in this case, in fact, there is

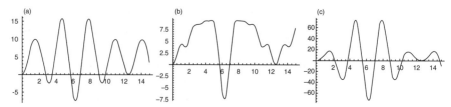

**Figure 8.2**   $\delta\Pi(t)$ for $n = 10$ and $(\omega(1), \omega(2)) = (1, 1)$ (a), $(\omega(1), \omega(2)) = (1, 10)$ (b), $(\omega(1), \omega(2)) = (10, 1)$ (c).

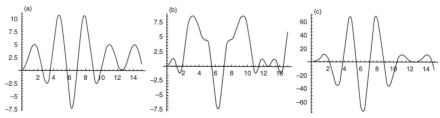

**Figure 8.3**   $\delta\Pi(t)$ for $n = 5$ and $(\omega(1), \omega(2)) = (1, 1)$ (a), $(\omega(1), \omega(2)) = (1, 10)$ (b), $(\omega(1), \omega(2)) = (10, 1)$ (c).

only a small range of time in which $\delta\Pi(t)$ is negative. In all the other intervals of time, $\delta\Pi(t)$ is positive and $\Pi(t)$ improves its original value. The situation is a bit less favorable for other choices of $(\omega(1), \omega(2))$. However, it should also be noticed that for these different choices of $\omega(1)$ and $\omega(2)$, $\tau$ can increase the value of his portfolio much more than in the other conditions. This is close to real life: you can choose an almost safe way to improve your portfolio for a large interval of time, or you can improve much more the value of $\delta\Pi(t)$ but taking the risk to *fall down in negative regions* (e.g., the interval $[2, 4]$ in Figure 8.2c). This result is due to the fact that as $\omega(1)$ and $\omega(2)$ are related to the initial values of the stock market, different initial conditions may correspond to quite different dynamical behaviors!

Now we change the relation between $M$ and $O$. Therefore, we fix $M = 2$, $O = 1$, $\lambda = 1$, $\omega_a = \omega_c = 1$, and $\Omega_A = \Omega_C = 2$. Again, the plots of $\delta\Pi(t)$ in Figure 8.4 are related to the following values: $(\omega(1), \omega(2)) = (1, 1), (1, 10), (10, 1)$, and we fix $n = 10$.

We see that these plots look very much like those in Figure 8.2 except that they are reflected with respect to the horizontal axis. This suggests that for $n = 10$, the main contribution in Equation 8.73 is the term $n(O - M)\sin^2(\lambda t)$. Of course, this is in agreement with our previous considerations, and suggests that, already for this value of $n$, the second and the third contributions in Equation 8.73 are much smaller than the first one.

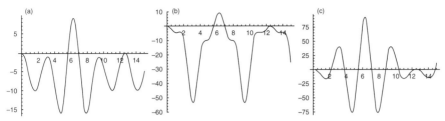

**Figure 8.4**  $\delta\Pi(t)$ for $n = 10$ $(\omega(1), \omega(2)) = (1, 1)$ (a), $(\omega(1), \omega(2)) = (1, 10)$ (b), $(\omega(1), \omega(2)) = (10, 1)$ (c).

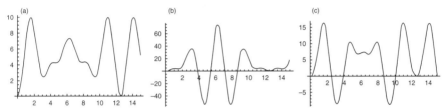

**Figure 8.5**  $\delta\Pi(t)$ for $n = 10$ $(\omega(1), \omega(2)) = (1, 1)$ (a), $(\omega(1), \omega(2)) = (1, 10)$ (b), $(\omega(1), \omega(2)) = (10, 1)$ (c).

We have already stressed that if $M = O$, then $\delta\Pi(t) = 0$ for all $t \geq 0$. Therefore, we do not plot $\delta\Pi(t)$ in this condition. Rather than this, we close this chapter considering what happens if we exchange the values of $\omega_a$ and $\omega_c$ with those of $\Omega_A$ and $\Omega_C$. For that we fix, as in Figure 8.2, $M = 1$, $O = 2$, $\lambda = 1$, whereas we take $\omega_a = \omega_c = 2$ and $\Omega_A = \Omega_C = 1$. The related plots are given in Figure 8.5.

This result is particularly interesting as it shows that if $\omega(1) = \omega(2) = 1$ and also for $n$ small enough, trader $\tau$ can only improve the value of his portfolio, no matter what the value of $t$ is. Remember now that $\omega(1) = \omega(2)$ is guaranteed, for instance, when the initial conditions for the trader $\tau$ and for the (only) trader of the reservoir, $\sigma$, coincide. Therefore, this suggests once more that the relation between the parameters $\omega_a, \omega_c$ and $\Omega_A, \Omega_C$, is crucial to determine $\Pi(t)$ and, in particular, that if we take $\Omega_A = \Omega_C < \omega_a = \omega_c$, $\tau$ is, in a certain sense, in a better condition with respect to trader $\sigma$, because its $\delta\Pi(t)$ is always positive. It is therefore natural to associate these parameters, as already claimed several times along these notes, to a sort of *information* reaching the traders: $\tau$ is *better informed* than $\sigma$. The quasiperiodic behavior of $\delta\Pi(t)$, which we expected to find for analytic reasons, seems to emerge from the plots here, and could be made more evident plotting $\delta\Pi(t)$ over a longer time interval.

# CHAPTER 9

# MODELS WITH AN EXTERNAL FIELD

In the previous chapter, a simple mechanism that fixes the price of the shares was proposed. However, as we have discussed, this strategy is not expected to be very useful in concrete applications as the price is chosen by the market supply, $\hat{O}$, and we have no general strategy to fix the value of $\hat{O}$. This means, of course, that we have no general strategy to fix the value of $\hat{P}$ as well. We only have a reasonable relation between $\hat{O}$ and $\hat{P}$, arising from the fact that the sum of these operators commute with $H$. As we have discussed, this relation could became relevant as soon as we are able to find $\langle\hat{O}\rangle$ by some means. As this possibility is not available yet, we adopt in this chapter a different point of view, which could be useful, in a more *evolved* model, to *tune* the coefficients of the Hamiltonian via a direct comparison with some experimental data. More in detail, the price of the share is not considered here as a *degree of freedom* of the model, but as an external known classical field needed in the definition of the interaction between the traders. On the other hand, we keep the other aspects of the general scheme discussed in the previous chapter. For this reason, the time-depending operator, $\hat{P}(t)$, the time dependence of which was deduced before using the Hamiltonian of the system, is replaced here by a classical function of time, $P(t)$, whose time dependence should be fixed, in principle, by the experimental data. It is clear that, in order to

*Quantum Dynamics for Classical Systems: With Applications of the Number Operator*,
First Edition. Fabio Bagarello.
© 2013 John Wiley & Sons, Inc. Published 2013 by John Wiley & Sons, Inc.

give a meaning to these experimental data, we should consider a somehow more realistic market. This can be achieved, first of all, considering different kinds of shares, as we do in Section 9.2. However, the evidence that we are still not so close to the description of a real stock market is given, for instance, by the absence of any financial derivative in our analysis.

## 9.1   THE MIXED MODEL

As in Section 8.3, we divide the SSM into two parts: the system $\mathcal{S}$, which consists of a single trader $\tau$, and the reservoir $\mathcal{R}$, which is made by all the traders of the market out of $\tau$. Also, in this section, we still consider a single type of shares, postponing the extension to different types of shares to Section 9.2. Adapting what we have discussed before, we consider here a *mixed model* containing a *Hamiltonian ingredient*, which involves the cash and the number of shares of both $\tau$ and of the traders in $\mathcal{R}$, and an *empirical part* because the dynamics of the price is not deduced as before from the market supply. On the contrary, the price is simply given as an input in Equation 8.58, which is replaced by the following system of differential equations

$$
\begin{cases}
\dfrac{d\hat{n}(t)}{dt} = i\lambda \left(-z^\dagger(t)\, Z(f,t) + z(t)\, Z^\dagger(\overline{f},t)\right), \\[2ex]
\dfrac{d\hat{k}(t)}{dt} = i\lambda\, P(t)\, \left(z^\dagger(t)\, Z(f,t) - z(t)\, Z^\dagger(\overline{f},t)\right), \\[2ex]
\dfrac{dz(t)}{dt} = i\left(P(t)\omega_c - \omega_a\right) z(t) + i\lambda\, Z(f,t)\, [z^\dagger(t), z(t)], \\[2ex]
\dfrac{dZ(f,t)}{dt} = i\left(P(t)\Omega_C - \Omega_A\right) Z(f,t) + i\lambda\, z(t)\, [Z^\dagger(\overline{f},t), Z(f,t)],
\end{cases}
$$

$$(9.1)$$

where $P(t)$ is a classical *external field*. In the rest of this section, we choose its analytical expression in four different simple ways to mimic different realistic time evolutions of the price of a share. These choices are also meant to simplify our computations as much as possible.

One could think of deducing system (Eq. 9.1) from the following effective (time-dependent) Hamiltonian $H_{\text{eff}}$, which looks very similar to the one in Equation 8.38:

$$
\begin{cases}
H_{\text{eff}}(t) = H_{\text{eff},0} + \lambda\, H_{\text{eff},I}(t), \quad \text{where} \\[1.5ex]
H_{\text{eff},0} = \omega_a\, \hat{n} + \omega_c\, \hat{k} + \sum_{k \in \Lambda}\left(\Omega_A\, \hat{N}_k + \Omega_C\, \hat{K}_k\right) \\[1.5ex]
H_{\text{eff}}(t) = \left(z^\dagger(t)\, Z(f,t) + z(t)\, Z^\dagger(\overline{f},t)\right)),
\end{cases}
$$

$$(9.2)$$

and where, for instance, $Z(f, t) = \sum_{k \in \Lambda} Z_k(t) f(k) = \sum_{k \in \Lambda} f(k) A_k C_k^{\dagger P(t)}$. However, this is not mathematically correct because, as we have discussed several times, the time dependence of $P(t)$ modifies the Heisenberg equation (Eq. 2.2) for the time evolution of a given operator $X$. Hence, it is more convenient to take Equation 9.1 as the starting point, and Equation 9.2 just as a formal Hamiltonian heuristically related to the economical system, and which will become relevant in the next section.

One of the problems, which makes the description of a stock market very hard, is to find a reasonable mechanism, which fixes the prices of all the shares. The reason this is not a trivial task is because we should be able to elaborate the output of the interactions between the various traders, as well as external sources of information reaching the traders. This, in principle, is exactly our final goal: wanting to be optimistic, we would like to produce a model, which contains $N$ traders and $L$ different kinds of shares whose prices are decided by some *global* mechanism to be identified. However, the analysis described in this part of the book has already shown that the preliminary models proposed so far give rise to absolutely non-trivial dynamical systems, which, in our opinion and independent of their possible economical meaning, deserve a deeper analysis already *per se*.

One of the main reasons we are interested in considering the model described by Equation 9.1 is to clarify as much as possible the role of the parameters appearing in that equation and in the Hamiltonians considered in Chapters 7 and 8. This is a very important step needed before undertaking any further generalization. We have already seen, for instance, that the parameters of the free Hamiltonians represent a sort of inertia or that their inverse can be seen as a sort of information reaching the traders. In this section, we try to justify further this interpretation. Hence, we play here with the parameters and the initial conditions. This analysis is also essential if we want to compare our approach with those already existing in the literature, where the role of certain parameters is quite often absolutely crucial, as in, for instance, Lux and Marchesi (1999) and Alfi et al. (2007), where the relations between the parameters of the models play a key role in determining the nature of the traders that form the market (Alfi et al., 2007) or the way in which these traders react to the dynamics of the market itself.

We do not use here a price function $P(t)$ deduced from experimental data, first of all because this might be rather irregular from an analytical point of view. Moreover, this choice does not make much sense here because of the unrealistic features of the model in Equation 9.1. Therefore, for the time being, we prefer to consider four different kinds of sufficiently regular functions $P_j(t)$, $j = 1, 2, 3, 4$, describing different possible behaviors of the price of the shares. More in detail, we consider the following forms for $P(t)$:

$$P_1(t) = t,$$

$$P_2(t) = \begin{cases} 0, & \text{if } t \in [0, 1] \\ t - 1, & \text{if } t \in [1, 3], \\ 2, & \text{if } t > 3, \end{cases}$$

$$P_3(t) = \begin{cases} 2, & \text{if } t \in [0, 1] \\ 3 - t, & \text{if } t \in [1, 3], \\ 0, & \text{if } t > 3, \end{cases}$$

$$P_4(t) = \begin{cases} 0, & \text{if } t \in [0, 1] \\ t - 1, & \text{if } t \in [1, 3] \\ 5 - t, & \text{if } t \in [3, 4] \\ 1, & \text{if } t > 4. \end{cases}$$

As we only perform numerical computations, we will fix here the time interval to be [0,6]. Hence, we are considering four different situations: $P_1(t)$ and $P_2(t)$ describe a price which, in [0,6], does not decrease, while $P_3(t)$ is a nonincreasing function. Finally, $P_4(t)$ describes a share whose value increases up to a maximum and then decreases and reach a limiting value. The reason for using these rather than other and more regular, for instance differentiable, functions of time is that with these choices it is easier to compute analytically some of the quantities appearing in the solution of system (Eq. 9.1).

Let us now show quickly how to find the solution of this system by means of the FPL approximation. For more details, we refer to Bagarello (2007). As in Section 8.3.2, the main idea is to find first an approximated solution for the last two equations of Equation 9.1 and then using these solutions in the first two equations of the same system. In this way, we recover the solution given in the previous chapter (Eq. 8.67), where $\eta_1(t)$, $\eta_2(t)$, $\tilde{\eta}_1(t)$, and $\tilde{\eta}_2(t)$ must be replaced by the following functions:

$$\begin{cases} \eta_1(t) = e^{i\chi(t)}, & \eta_2(t) = i\lambda \int_0^t e^{i\tilde{\chi}(t')}\, dt' \\ \tilde{\eta}_1(t) = e^{i\tilde{\chi}(t)}, & \tilde{\eta}_2(t) = i\lambda \int_0^t e^{i\chi(t')}\, dt', \end{cases} \tag{9.3}$$

with   $\chi(t) = \int_0^t (P(t')\omega_c - \omega_a)\,dt'$   and   $\tilde{\chi}(t) = \int_0^t (P(t')\Omega_C - \Omega_A)\,dt'$.
These results can be used in the first two equations in (9.1) and we get

$$\begin{cases} n(t) = n + \delta n(t) = n - 2\lambda \int_0^t \Im(r(t'))\,dt' \\ k(t) = k + \delta k(t) = k + 2\lambda \int_0^t P(t')\,\Im(r(t'))\,dt', \end{cases} \tag{9.4}$$

with obvious notation, having introduced the function $r(t)$ as in
Equation 8.70. The time evolution of the portfolio of $\tau$ is $\Pi(t) = \Pi(0) + \delta\Pi(t)$, and we get

$$\delta\Pi(t) = n(P(t) - P(0)) + P(t)\,\delta n(t) + \delta k(t). \tag{9.5}$$

As in Section 8.3.2, to make our analysis easier, we consider here $\mathcal{R}$ as
consisting in just a single trader, $\sigma$, provided with $n'$ shares and $k'$ units of
cash at $t = 0$. This is not a major requirement. In fact, the reservoir $\mathcal{R}$ is,
in our present approach, a single *entity* that could be made of many, few,
or very few traders, without any change in the analytical results obtained
so far. This choice makes the computation of $\omega(1)$ and $\omega(2)$ in Equation
8.70 rather easy and we get

$$\begin{cases} \omega(2) = -|f(1)|^2(1+n')\dfrac{k!}{(k-M)!}\dfrac{k'!}{(k'-M)!} \\ \omega(1) = \omega(2) + |f(1)|^2 n'\dfrac{k!}{(k-M)!}\dfrac{(k'+M)!}{k'!}, \end{cases} \tag{9.6}$$

which depends, not surprisingly, on the initial values of the market. In
particular, as in Chapter 8, $M$ is the price of the share at $t = 0$. As a
second assumption, we just consider the case of a trader $\tau$, which is
*entering into the market*, so that it possesses no share at all at $t = 0$.
Therefore, we take $n = 0$. This has an immediate consequence: in order
to have an economical meaning, $\delta n(t)$ can only be nonnegative! This is
actually what happens in all the explicit computations we have considered
so far.

Let us now analyze the numerical results of the solutions of system
(Eq. 9.1) for the different choices of $P(t)$ introduced previously. Here,
for concreteness' sake, we fix $f(1) = 10^{-3}$ in Equation 9.6 and consider the following *cases* (different choices of the parameters of the free
Hamiltonian, $H_0$) and *subcases* (different initial conditions):

**Case I:** $(\omega_a, \omega_c, \Omega_A, \Omega_C) = (1, 1, 1, 1)$;
**Case II:** $(\omega_a, \omega_c, \Omega_A, \Omega_C) = (10, 10, 1, 1)$;
**Case III:** $(\omega_a, \omega_c, \Omega_A, \Omega_C) = (1, 1, 10, 10)$;

**Case IV:** $(\omega_a, \omega_c, \Omega_A, \Omega_C) = (20, 10, 5, 1)$;
**Case V:** $(\omega_a, \omega_c, \Omega_A, \Omega_C) = (1, 5, 10, 20)$;
**Case VI:** $(\omega_a, \omega_c, \Omega_A, \Omega_C) = (1, 3, 2, 7)$.
**Subcase a:** $(k, k', n', M) = (20, 20, 100, 2)$;
**Subcase b:** $(k, k', n', M) = (80, 20, 100, 2)$;
**Subcase c:** $(k, k', n', M) = (20, 80, 100, 2)$;
**Subcase d:** $(k, k', n', M) = (80, 80, 100, 2)$.

We also recall here that $n = 0$ everywhere in this analysis.[1] The first interesting result of our numerical computations is that there are several situations, such as IIIa, Va, and VIa for $P_1(t)$, or IIIa, IVa, Va, and VIa for $P_2(t)$, for which the variation of $\delta n$ is larger than 0 but always (i.e., for $0 < t < 6$) strictly smaller than 1. Analogously, there are other situations in which $\delta k(t)$, belongs to the interval $[-1, 1]$ for $0 < t < 6$, such as in IIIa, Va, and VIa for $P_1(t)$. We refer to Bagarello (2009) for more details. This can be interpreted as follows: as there is not much cash being exchanged in the market, it is quite unlikely that some transaction takes place between the two traders $\tau$ and $\sigma$: $\tau$ wants to buy a share but the transaction cannot occur! Figure 9.1 shows $\delta n(t)$ and $\delta k(t)$ for $P_1(t)$ in Case Va.

Moreover, there are situations in which the total quantity of cash or the total number of shares exceed their initial values, such as in Case Id or IId for $P_2(t)$ (Figure 9.2). This is clearly incompatible with the fact that these operators should be constant as ours is a closed market,[2]

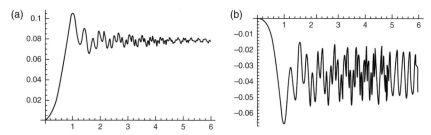

**Figure 9.1** $\delta n(t)$ (a) and $\delta k(t)$ (b) for $P_1(t)$, Case Va: no transaction is possible for $t \in [0, 6]$.

---

[1] As, to avoid dealing with too many changing parameters, we are only considering $M = 2$ in our analysis, what follows is valid, in principle, only for $P_3(t)$. However, adding the constant value 2 in the definition of the other $P_j(t)$, $j = 1, 2, 4$, we go back to $M = 2$. Not many differences are expected to arise.

[2] Notice that as we have no Hamiltonian here associated with system (Eq. 9.1), we cannot conclude that the above number operators commute with the Hamiltonian, as we claimed several times along these notes. In addition, **R1–R5** have no meaning here.

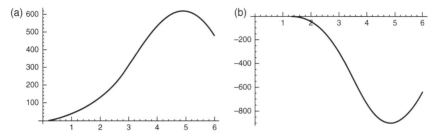

**Figure 9.2**    $\delta n(t)$, Case Id (a), and $\delta k$, Case IId (b), for $P_2(t)$: the FPL approximation does not hold in all of $[0, 6]$.

and it is uniquely due to the FPL approximation, which we have adopted to solve the original system of differential equations. In other words, in these cases, there exists a value of $t$, $t_f < 6$, such that for $t > t_f$, the FPL approximation does not work and our results become meaningless. This happens usually in Subcase d, because $\omega(1)$ and $\omega(2)$ assume very high values as a consequence of the initial conditions, which, in this particular subcase, are large integer numbers. There exists a trivial way to avoid these *explosions*: it is enough to choose a smaller value for $f(1)$. However, this new choice would create more problems of the opposite nature: no transaction at all is possible for *small* initial conditions, that is, for Subcase a. In other words, using our approximation and for fixed initial conditions, $f(1)$ behaves like a threshold value: if $f(1)$ is too small, a large amount of cash and a large number of shares are needed for some transactions to take place. On the other hand, if $f(1)$ is too large, then the FPL approximation breaks down very soon. This problem could be solved, of course, if we adopt a better approximation to solve the differential equation in (9.1). This is a part of the work in progress, but is far from being an easy task, also because of the noncommutative nature of the variables appearing in the system to be solved. This difficulty is increased by another, nonnegligible, feature of the system (Eq. 9.1), which is not linear.

We also observe that the range of variations of all the relevant observables, that is, $\delta n(t)$, $\delta k(t)$, and $\delta \Pi(t)$, is minimum in Subcase a, maximum in Subcase d, and of the same order of magnitude in Subcases b and c. This is again due to the fact that different initial conditions produce significantly different values of $\omega(1)$ and $\omega(2)$, which increase very quickly with $k$ and $k'$; see Equation 9.6. We further observe that the analytic behavior of $\delta \Pi(t)$ is close (sometimes very close) to that of $P(t)$. This is shown in Figure 9.3, which is obtained by choosing $P(t) = P_4(t)$: we plot $P_4(t)$ in (a), while $\delta \Pi(t)$ is shown in Cases Ic (b), IVd (c), and Vc (d). From these figures, it is clear that when the price of the shares stays constant, there is no variation of $\delta\Pi(t)$. On the contrary, this increases when $P(t)$ increases and decreases when $P(t)$ decreases. This is not related to the particular choice of $P(t)$ but can be analytically deduced from formula (9.5) with $n = 0$, which can be written as $\delta\Pi(t) = P(t)\delta n(t) + \delta k(t)$. Indeed, if we just compute the time derivative of both sides and use the definitions of $\delta n(t)$ and $\delta k(t)$, we find that $\dot\Pi(t) = \dot P(t)n(t)$. Then, as $n(t)$ is not negative, $\dot\Pi(t)$ and $\dot P(t)$ have the same sign. More than this, when $P(t)$ is constant, $\Pi(t)$ is also constant. Figure 9.3 also shows that there are conditions (e.g., Vc), in which $\delta\Pi(t)$ looks very much the same as $P(t)$, while with different choices of the constants of the Hamiltonian and of the initial conditions, the shape of $\delta\Pi(t)$ appears sufficiently different from that of $P(t)$, even if it still reflects our previous considerations.

### 9.1.1   Interpretation of the Parameters

As in all previous models, a special role is played by the parameters of the free Hamiltonian. Hence, we close this section by focusing our attention on the variation of the portfolio of $\tau$ as this might give hints on what these parameters are useful for. Actually, this was really our original motivation to consider system (Eq. 9.1). This analysis is based on Table 9.1, where we list, approximately, the range of variations of $\delta\Pi(t)$ for the different cases, subcases, and choices of $P(t)$.

We see that, for each fixed subcase, Case I is the one in which we find the largest variation of $\delta\Pi(t)$. In particular, if we consider $P(t) = P_j(t)$, with $j = 1, 2, 4$, $\delta\Pi(t)$ increases its value more than that in Cases II–VI for any fixed subcase. If $j = 3$, then $\delta\Pi(t)$ decreases its value in Case I more than that in Cases II–VI. This difference between $P_3(t)$ and the others, $P_j(t)$, $j \neq 3$, is clearly related to our previous remark

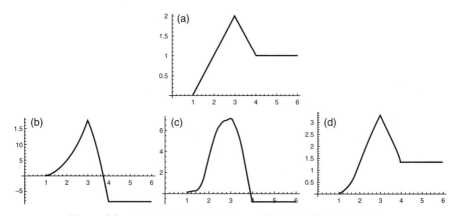

**Figure 9.3**   (a) $P_4(t)$. $\delta\Pi(t)$ in Cases (b) Ic, (c) IVd, and (c) Vc.

**TABLE 9.1  Range of Variation of $\delta\Pi(t)$ for Different Choices of $P_j(t)$**

|          | $P_1(t)$          | $P_2(t)$           | $P_3(t)$           | $P_4(t)$            |
|----------|-------------------|--------------------|--------------------|---------------------|
| **I** a  | From 0 to 4       | From 0 to 1.2      | From −1.2 to 0     | From −0.5 to 1.25   |
| **I** b  | From 0 to 60      | From 0 to 17.5     | From −17.5 to 0    | From −7 to 17       |
| **I** c  | From 0 to 70      | From 0 to 20       | From −20 to 0      | From −8 to 20       |
| **I** d  | From 0 to 900     | From 0 to 300      | From −300 to 0     | From −110 to 300    |
| **II** a | From 0 to 3.5     | From 0 to 1        | From −1 to 0       | From −0.4 to 1      |
| **II** b | From 0 to 55      | From 0 to 17       | From −16 to 0      | From −7 to 17       |
| **II** c | From 0 to 55      | From 0 to 17.5     | From −17.5 to 0    | From −7 to 17       |
| **II** d | From 0 to 900     | From 0 to 270      | From −280 to 0     | From −100 to 250    |
| **III** a| From 0 to 1       | From 0 to 0.35     | From −0.3 to 0     | From −0.1 to 0.3    |
| **III** b| From 0 to 8       | From 0 to 2.5      | From −2.5 to 0     | From −0.5 to 2.5    |
| **III** c| From 0 to 17.5    | From 0 to 5.2      | From −5.2 to 0     | From −2 to 5.3      |
| **III** d| From 0 to 140     | From 0 to 43       | From −43 to 0      | From −10 to 45      |
| **IV** a | From 0 to 0.7     | From 0 to 0.025    | From −0.15 to 0    | From −0.01 to 0.03  |
| **IV** b | From 0 to 9       | From 0 to 0.4      | From −1.4 to 0     | From −0.05 to 0.4   |
| **IV** c | From 0 to 10      | From 0 to 0.45     | From −2.5 to 0     | From −0.2 to 0.42   |
| **IV** d | From 0 to 150     | From 0 to 7        | From −22 to 0      | From −0.4 to 7      |
| **V** a  | From 0 to 0.4     | From 0 to 0.2      | From −0.03 to 0    | From 0 to 0.2       |
| **V** b  | From 0 to 6       | From 0 to 1.25     | From −0.4 to 0     | From 0 to 1.3       |
| **V** c  | From 0 to 7       | From 0 to 3.2      | From −0.5 to 0     | From 0 to 3.5       |
| **V** d  | From 0 to 100     | From 0 to 22       | From −7 to 0       | From 0 to 22        |
| **VI** a | From 0 to 0.6     | From 0 to 0.7      | From −0.06 to 0    | From 0 to 0.7       |
| **VI** b | From 0 to 8       | From 0 to 9        | From −0.6 to 0     | From 0 to 9         |
| **VI** c | From 0 to 10      | From 0 to 12       | From −1 to 0       | From 0 to 12        |
| **VI** d | From 0 to 120     | From 0 to 150      | From −11 to 0      | From 0 to 150       |

concerning the signs of $\delta\dot\Pi(t)$ and $\dot P(t)$. More explicitly, whereas $P_3(t)$ is a nonincreasing function, and therefore $\dot P_3(t) \le 0$, the time derivatives of the other price functions are always larger or equal to zero.

If we now compare Cases II and III, considering equal subcases, we see that $\tau$ is favored for the first choice of constants more than for the second. In fact, considering, for instance, $P_1(t)$, while in Case IIa, $\delta\Pi(t)$ starts from 0 and reaches the value 3.5, in Case IIIa, $\delta\Pi(t)$ assumes values between 0 and 1. Analogously, while in IIc, $\delta\Pi(t) \in [0, 55]$, in IIIc, $\delta\Pi(t) \in [0, 17.5]$. Similar conclusions are deduced also for the other $P_j(t)$. Therefore, at the first sight, we could try to interpret in the present settings $\omega_a$, $\omega_c$, $\Omega_A$, and $\Omega_C$ (i.e., those parameters entering in the definition of the cases) as related to the information reaching, respectively, $\tau$ and $\sigma$: when $\tau$ gets a larger amount of information, he is able to earn more! As a matter of fact, our previous analysis suggests that what is really important is not the single parameter $\omega_a$ (or $\Omega_A$) but rather their differences $\omega_a - \omega_c$ (or $\Omega_A - \Omega_C$). The same conclusion can also be deduced here: see, for instance, Case IV, in which $\omega_a$ and $\omega_c$ are different (and larger than $\Omega_A$ and $\Omega_C$). In this condition, we see from Table 9.1 that $\tau$ does not improve his portfolio as in Cases I, II, and III: so it seems quite plausible that what is really relevant is some suitable relation between $\omega_a$ and $\omega_b$, and between $\Omega_A$ and $\Omega_B$, rather than the values of the constants themselves. However, this is not enough: our numerical results clearly show that there exists some asymmetry between $(\omega_a, \omega_c)$ and $(\Omega_A, \Omega_C)$; the $\omega$s seem to carry a higher amount of information than that of the $\Omega$s. The reason for this asymmetry stands probably in the difference between $\mathcal{S}$ and $\mathcal{R}$, that is, by the different ways in which $\tau$ and $\sigma$ are described in Equation (9.1).

Once again, the role of the parameters of $H_0$ turns out to be crucial in determining the time evolution of the interacting system. Therefore, a good tuning of these parameters may be useful to get a better fit of the experimental data.

Of course, it would be quite interesting to improve the FPL approximation that we have adopted here in order to avoid the spurious nonconservation of the integrals of motion observed in our numerical computations. This is part of our future plans. On the other hand, we do not expect that changing the values of $M$ in the definition of our subcases may lead to different conclusions from those outlined here.

## 9.2   A TIME-DEPENDENT POINT OF VIEW

In this section, we introduce several kinds of shares and adopt a slightly different point of view. Our Hamiltonian has no $H_{price}$ contribution at all as the price operators $\hat{P}_\alpha$, $\alpha = 1, \ldots, L$, are now replaced from the very beginning, as in the previous section, by external classical fields $P_\alpha(t)$, whose time dependence describes, as an input of the model, the variation of the prices of the shares. This implies that any change of the prices is automatically included in the model through the analytical expressions of the functions $P_\alpha(t)$. Hence, the interaction Hamiltonian $H_I$ turns out to be a time-dependent operator, $H_I(t)$. More in detail, the Hamiltonian is $H(t) = H_0 + \lambda H_I(t)$, which we can write by introducing the time-dependent *selling* and *buying* operators that extend those introduced in Chapter 8,

$$x_{j,\alpha}(t) := a_{j,\alpha} c_j^{\dagger^{P_\alpha(t)}}, \qquad x_{j,\alpha}^\dagger(t) := a_{j,\alpha}^\dagger c_j^{P_\alpha(t)}, \qquad (9.7)$$

as

$$H(t) = \sum_{j,\alpha} \omega_{j,\alpha} \hat{n}_{j,\alpha} + \sum_j \omega_j \hat{k}_j + 2\lambda \sum_{i,j,\alpha} p_{i,j}^{(\alpha)} x_{i,\alpha}^\dagger(t) x_{j,\alpha}(t), \quad (9.8)$$

where $H_{\mathrm{I}}(t) = 2 \sum_{i,j,\alpha} p_{i,j}^{(\alpha)} x_{i,\alpha}^{\dagger}(t)\, x_{j,\alpha}(t)$ and $H_0 = \sum_{j,\alpha} \omega_{j,\alpha}\, \hat{n}_{j,\alpha} +$ $\sum_j \omega_j \hat{k}_j$. This is very close to what we have done in the previous sections, except for the essential fact that $H$ in Equation 9.8 is a *true*, self-adjoint, time-dependent Hamiltonian, rather than an effective Hamiltonian deduced a posteriori from the system of differential equation in (9.1). We stress once more that neither $H(t)$ in Equation 9.8 nor $H_{\mathrm{eff}}$ in Equation 9.2 produces the system equation in (9.1).

Concerning the notation, while the Latin indices are related to the traders, the Greek ones appear here because of the different kinds of shares considered in this section: more than a single kind of shares exist (finally!) in our SSM. The related $L$ price function $P_\alpha(t)$ is taken as the piecewise constant as it is quite natural to assume that the price of a share changes discontinuously: it has a certain value before two transactions and (in general) a different value after the transaction. Again, this new value does not change until the next transaction takes place. More in detail, we introduce a time step $h$ that we call *the time of transaction*, and we divide the interval $[0, t[$ in subintervals which, for simplicity, we consider having the same duration $h$: $[0, t[= [t_0, t_1[\cup[t_0, t_1[\cup[t_1, t_2[\cdots[t_{M-1}, t_M[$, where $t_0 = 0, t_1 = h, \ldots, t_{M-1} = (M-1)h = t - h, t_M = Mh = t$. Hence, $h = t/M$. As for the prices, we put

$$P_\alpha(t) = \begin{cases} P_{\alpha,0}, & t \in [t_0, t_1], \\ P_{\alpha,1}, & t \in [t_1, t_2], \\ \ldots\ldots, & \\ P_{\alpha,M-1}, & t \in [t_{M-1}, t_M], \end{cases} \tag{9.9}$$

for $\alpha = 1, \ldots L$. An o.n. basis in the Hilbert space $\mathcal{H}$ of the model is now the set of vectors defined as

$$\varphi_{\{n_{j,\alpha}\};\{k_j\}} := \frac{a_{1,1}^{\dagger\; n_{1,1}} \cdots a_{N,L}^{\dagger\; n_{N,L}} c_1^{\dagger k_1} \cdots c_N^{\dagger\; k_N}}{\sqrt{n_{1,1}! \cdots n_{N,L}! k_1! \cdots k_L!}} \varphi_0, \tag{9.10}$$

where $\varphi_0$ is the vacuum of all the annihilation operators involved here. To simplify the notation, we introduce a set $\mathcal{F} = \{\{n_{j,\alpha}\}; \{k_j\}\}$ so that the vectors of the basis are simply written as $\varphi_{\mathcal{F}}$. The difference between these and the vectors in Equation 8.4 stands clearly in the absence of the *quantum numbers* $O$ and $M$. This is clearly due to the fact that $P(t)$ is no longer a degree of freedom of the system, and that the market supply does not even exist anymore, here.

The main problem we want to discuss now is the following: suppose that at $t = 0$, the market is described by a vector $\varphi_{\mathcal{F}_0}$. This means that as

$\mathcal{F}_0 = \{\{n^o_{j,\alpha}\}, \{k^o_j\}\}$, at $t = 0$, the trader $\tau_1$ has $n^o_{1,1}$ shares of the kind $\Sigma_1$, $n^o_{1,2}$ shares of $\Sigma_2$, ..., and $k^o_1$ units of cash. Analogously, the trader $\tau_2$ has $n^o_{2,1}$ shares of $\Sigma_1$, $n^o_{2,2}$ shares of $\Sigma_2$, ..., and $k^o_2$ units of cash, and so on. We want to compute the probability that at time $t$, the market has moved to the configuration $\mathcal{F}_f = \{\{n^f_{j,\alpha}\}, \{k^f_j\}\}$. This means that, for example, $\tau_1$ has now $n^f_{1,1}$ shares of $\Sigma_1$, $n^f_{1,2}$ shares of $\Sigma_2$, ..., and $k^f_1$ units of cash.

Similar problems arise quite often in ordinary quantum mechanics: we need to compute a probability transition from the original state $\varphi_{\mathcal{F}_0}$ to a final state $\varphi_{\mathcal{F}_f}$, and this is the reason we use here a somehow standard time-dependent perturbation scheme for which we refer to Messiah (1962). The main difference with respect to what we have done in the previous section is that we use the Schrödinger rather than the Heisenberg picture. Hence, the market is described by a time-dependent wave function $\Psi(t)$ that, for $t = 0$, reduces to $\varphi_{\mathcal{F}_0}$: $\Psi(0) = \varphi_{\mathcal{F}_0}$. The transition probability we are looking for is

$$P_{\mathcal{F}_0 \to \mathcal{F}_f}(t) := \left| < \varphi_{\mathcal{F}_f}, \Psi(t) > \right|^2. \tag{9.11}$$

The computation of $P_{\mathcal{F}_0 \to \mathcal{F}_f}(t)$ is a standard exercise; see Messiah (1962) and Section 2.5.1. We give here the main steps of its derivation.

As the set of vectors $\varphi_{\mathcal{F}}$ is an o.n. basis in $\mathcal{H}$, the wave function $\Psi(t)$ can be written as

$$\Psi(t) = \sum_{\mathcal{F}} c_{\mathcal{F}}(t) \, e^{-iE_{\mathcal{F}}t} \varphi_{\mathcal{F}}, \tag{9.12}$$

where $E_{\mathcal{F}}$ is the eigenvalue of $H_0$ defined as

$$H_0 \varphi_{\mathcal{F}} = E_{\mathcal{F}} \varphi_{\mathcal{F}}, \quad \text{where} \quad E_{\mathcal{F}} = \sum_{j,\alpha} \omega_{j,\alpha} n_{j,\alpha} + \sum_j \omega_j k_j. \tag{9.13}$$

This is a consequence of the fact that $\varphi_{\mathcal{F}}$ in Equation 9.10 is an eigenstate of $H_0$. As in Section 2.5.1, we call $E_{\mathcal{F}}$ the free energy of $\varphi_{\mathcal{F}}$. Putting Equation 9.12 in Equation 9.11 and recalling that if the eigenstates are not degenerate,[3] the corresponding eigenvectors are orthogonal, $< \varphi_{\mathcal{F}}, \varphi_{\mathcal{G}} > = \delta_{\mathcal{F},\mathcal{G}}$, we have

$$P_{\mathcal{F}_0 \to \mathcal{F}_f}(t) := \left| c_{\mathcal{F}_f}(t) \right|^2 \tag{9.14}$$

---

[3]This is assumed here: it is just a matter of choice of the parameters of the free Hamiltonian.

The answer to our original question is, therefore, given if we are able to compute $c_{\mathcal{F}_f}(t)$ in Equation 9.12. Owing to the analytic form of our Hamiltonian, this cannot be done exactly. Following Section 2.5.1, we adopt here a simple perturbation expansion in the interaction parameter $\lambda$ appearing in the Hamiltonian $H$ in Equation 9.8. In other words, we look for the coefficients in Equation 9.12 having the form

$$c_{\mathcal{F}}(t) = c_{\mathcal{F}}^{(0)}(t) + \lambda c_{\mathcal{F}}^{(1)}(t) + \lambda^2 c_{\mathcal{F}}^{(2)}(t) + \cdots \tag{9.15}$$

Each $c_{\mathcal{F}}^{(j)}(t)$ satisfies a differential equation that has been deduced previously; see Equation 2.43:

$$\begin{cases} \dot{c}_{\mathcal{F}'}^{(0)}(t) = 0, \\ \dot{c}_{\mathcal{F}'}^{(1)}(t) = -i \sum_{\mathcal{F}} c_{\mathcal{F}}^{(0)}(t)\, e^{i(E_{\mathcal{F}'} - E_{\mathcal{F}})t} < \varphi_{\mathcal{F}'}, H_I(t)\varphi_{\mathcal{F}} >, \\ \dot{c}_{\mathcal{F}'}^{(2)}(t) = -i \sum_{\mathcal{F}} c_{\mathcal{F}}^{(1)}(t)\, e^{i(E_{\mathcal{F}'} - E_{\mathcal{F}})t} < \varphi_{\mathcal{F}'}, H_I(t)\varphi_{\mathcal{F}} >, \\ \cdots \cdots \cdots , \end{cases} \tag{9.16}$$

The first equation, together with the initial condition $\Psi(0) = \varphi_{\mathcal{F}_0}$, gives $c_{\mathcal{F}'}^{(0)}(t) = c_{\mathcal{F}'}^{(0)}(0) = \delta_{\mathcal{F}', \mathcal{F}_0}$. When we replace this solution in the differential equation for $c_{\mathcal{F}'}^{(1)}(t)$ we get, recalling again that $\Psi(0) = \varphi_{\mathcal{F}_0}$,

$$c_{\mathcal{F}'}^{(1)}(t) = -i \int_0^t e^{i(E_{\mathcal{F}'} - E_{\mathcal{F}_0})t_1} < \varphi_{\mathcal{F}'}, H_I(t_1)\varphi_{\mathcal{F}_0} > dt_1, \tag{9.17}$$

at least if $\mathcal{F}_0 \neq \mathcal{F}'$. Using this in Equation 9.16, we further get

$$c_{\mathcal{F}'}^{(2)}(t) = (-i)^2 \sum_{\mathcal{F}} \int_0^t \left( \int_0^{t_2} e^{i(E_{\mathcal{F}} - E_{\mathcal{F}_0})t_1}\, h_{\mathcal{F}, \mathcal{F}_0}(t_1)\, dt_1 \right)$$

$$e^{i(E_{\mathcal{F}'} - E_{\mathcal{F}})t_2}\, h_{\mathcal{F}', \mathcal{F}}(t_2)\, dt_2, \tag{9.18}$$

where we have introduced the shorthand notation

$$h_{\mathcal{F}, \mathcal{G}}(t) := < \varphi_{\mathcal{F}}, H_I(t)\varphi_{\mathcal{G}} > . \tag{9.19}$$

Of course, higher-order corrections could also be deduced simply by iterating this procedure.

## 9.2.1  First-Order Corrections

We adapt now this general procedure to the analysis of our stock market, by computing $P_{\mathcal{F}_0 \to \mathcal{F}_f}(t)$ in Equation 9.14 up to the first-order corrections in powers of $\lambda$ and assuming that $\mathcal{F}_f$ is different from $\mathcal{F}_0$. Hence, we have

$$P_{\mathcal{F}_0 \to \mathcal{F}_f}(t) = \left| c_{\mathcal{F}_f}^{(1)}(t) \right|^2 = \lambda^2 \left| \int_0^t e^{i(E_{\mathcal{F}_f} - E_{\mathcal{F}_0})t_1} h_{\mathcal{F}_f, \mathcal{F}_0}(t_1)\, dt_1 \right|^2. \quad (9.20)$$

Using Equation 9.9 and introducing $\delta E = E_{\mathcal{F}_f} - E_{\mathcal{F}_0}$, after some simple computations we get

$$P_{\mathcal{F}_0 \to \mathcal{F}_f}(t) = \lambda^2 \left( \frac{\delta E\, h/2}{\delta E/2} \right)^2 \left| \sum_{k=0}^{M-1} h_{\mathcal{F}_f, \mathcal{F}_0}(t_k)\, e^{it_k \delta E} \right|^2. \quad (9.21)$$

The matrix elements $h_{\mathcal{F}_f, \mathcal{F}_0}(t_k)$ can be easily computed. Indeed, because of some standard properties of the bosonic operators, we find that

$$a_{i,\alpha}^\dagger a_{j,\alpha} c_i^{P_{\alpha,k}} c_j^{\dagger P_{\alpha,k}} \varphi_{\mathcal{F}_0} = \Gamma_{i,j;\alpha}^{(k)} \varphi_{\mathcal{F}_{0,k}^{(i,j,\alpha)}},$$

where

$$\Gamma_{i,j;\alpha}^{(k)} := \sqrt{\frac{(k_j^o + P_{\alpha,k})!}{k_j^o!} \frac{k_i^o!}{(k_i^o - P_{\alpha,k})!}}\, n_{j,\alpha}^o\,(1 + n_{i,\alpha}^o), \quad (9.22)$$

and $\mathcal{F}_{0,k}^{(i,j,\alpha)}$ differs from $\mathcal{F}_0$ only for the following replacements: $n_{j,\alpha}^o \to n_{j,\alpha}^o - 1$, $n_{i,\alpha}^o \to n_{i,\alpha}^o + 1$, $k_j^o \to k_j^o + P_{\alpha,k}$, $k_i^o \to k_i^o - P_{\alpha,k}$. Note that in our computations, we are implicitly assuming that $k_i^o \geq P_{\alpha,k}$, for all $i$, $k$, and $\alpha$. This is because otherwise the trader $\tau_i$ would have not enough money to buy a share $\Sigma_\alpha$. A similar problem was already discussed in Section 8.2.

We find that

$$h_{\mathcal{F}_f, \mathcal{F}_0}(t_k) = 2 \sum_{i,j,\alpha} p_{i,j}^{(\alpha)} \Gamma_{i,j;\alpha}^{(k)} < \varphi_{\mathcal{F}_f}, \varphi_{\mathcal{F}_{0,k}^{(i,j,\alpha)}} > . \quad (9.23)$$

Of course, due to the orthogonality of the vectors $\varphi_{\mathcal{F}}$s, the scalar product $< \varphi_{\mathcal{F}_f}, \varphi_{\mathcal{F}_{0,k}^{(i,j,\alpha)}} >$ is different from zero (and equal to one) if and only if $n_{j,\alpha}^f = n_{j,\alpha}^o - 1$, $n_{i,\alpha}^f = n_{i,\alpha}^o + 1$, $k_i^f = k_i^o - P_{\alpha,k}$, and $k_j^f = k_j^o + P_{\alpha,k}$, and, moreover, if all the other *new* and *old* quantum numbers coincide.

For concreteness sake, we now consider two simple situations: in the first example later, we just assume that the prices of the various shares do not change with $t$. In the second example, we consider the case in which only few changes occur. For concreteness, we fix $M = 3$.

**Example 9.1  Constant Prices**  *Let us assume that for all $k$ and for all $\alpha$, $P_{\alpha,k} = P_\alpha(t_k) = P_\alpha$. This means that $\Gamma_{i,j;\alpha}^{(k)}$, $\mathcal{F}_{0,k}^{(i,j,\alpha)}$, and the related vectors $\varphi_{\mathcal{F}_{0,k}^{(i,j,\alpha)}}$ do not depend on $k$. Therefore, $h_{\mathcal{F}_f,\mathcal{F}_0}(t_k)$ is also independent of $k$. After few computations, we get*

$$P_{\mathcal{F}_0 \to \mathcal{F}_f}(t) = \lambda^2 \left( \frac{\sin(\delta E t/2)}{\delta E/2} \right)^2 \left| h_{\mathcal{F}_f,\mathcal{F}_0}(0) \right|^2 \qquad (9.24)$$

*to which corresponds the following transition probability per unit of time:*

$$p_{\mathcal{F}_0 \to \mathcal{F}_f} = \lim_{t,\infty} \frac{1}{t} P_{\mathcal{F}_0 \to \mathcal{F}_f}(t) = 2\pi \lambda^2 \delta(E_{\mathcal{F}_f} - E_{\mathcal{F}_0}) \left| h_{\mathcal{F}_f,\mathcal{F}_0}(0) \right|^2. \qquad (9.25)$$

*This formula shows that in this limit, and in our approximations, a transition between the states $\varphi_{\mathcal{F}_0}$ and $\varphi_{\mathcal{F}_f}$ is possible only if the two states have the same free energy. The presence of $h_{\mathcal{F}_f,\mathcal{F}_0}(0)$ in formulas (9.24) and (9.25) shows, using our previous remark, that at the order we are working here, a transition is possible only if $\varphi_{\mathcal{F}_0}$ does not differ from $\varphi_{\mathcal{F}_f}$ for more than one share in two of the $n_{j,\alpha}s$ and for more than $P_\alpha$ units of cash in two of the $k_j s$.[4] All the other transitions, for instance, those in which the numbers of shares differ for more than one unit, are forbidden to this order in perturbation theory.*

**Example 9.2  Few Changes in the Price**  *Let us now fix $M = 3$. Formula (9.21) can be rewritten as*

$$P_{\mathcal{F}_0 \to \mathcal{F}_f}(t) = 4\lambda^2 \left( \frac{\sin(\delta E h/2)}{\delta E/2} \right)^2 \Big| \sum_{i,j,\alpha} p_{i,j}^{(\alpha)} (\Gamma_{i,j;\alpha}^{(0)} < \varphi_{\mathcal{F}_f}, \varphi_{\mathcal{F}_{0,0}^{(i,j,\alpha)}} > +$$

$$+ \Gamma_{i,j;\alpha}^{(1)} < \varphi_{\mathcal{F}_f}, \varphi_{\mathcal{F}_{0,1}^{(i,j,\alpha)}} > \mathrm{e}^{\mathrm{i}h\delta E} + \Gamma_{i,j;\alpha}^{(2)} < \varphi_{\mathcal{F}_f}, \varphi_{\mathcal{F}_{0,2}^{(i,j,\alpha)}} > \mathrm{e}^{2\mathrm{i}h\delta E}) \Big|^2 \qquad (9.26)$$

---

[4]Of course, these differences must involve just two traders. Therefore, they must be related to just two values of $j$.

*The meaning of this formula is not very different from the one deduced in the previous example: if we work at this order of approximation, the only possibilities for a transition $\mathcal{F}_0 \to \mathcal{F}_f$ to occur are essentially those already discussed in Example 9.1. The difference is that because of the different values of the price of the share, just one scalar product (if any) in Equation 9.26 needs to be different from zero to get a nontrivial value of the transition probability. It is not surprising that in order to get something different, we need to go to higher orders in powers of $\lambda$.*

Concerning the validity of the approximation, let us consider the easiest situation: we have constant prices (Example 9.1) and, moreover, in the summations in Equation 9.23, only one contribution survives, the one with $i_0$, $j_0$, and $\alpha_0$. Then, we have

$$h_{\mathcal{F}_f, \mathcal{F}_0}(t_0) = 2 p_{i_0, j_0}^{(\alpha_0)} \Gamma_{i_0, j_0; \alpha_0} < \varphi_{\mathcal{F}_f}, \varphi_{\mathcal{F}_0^{(i_0, j_0, \alpha_0)}} > .$$

Because of Equation 9.24 and as our approximation becomes meaningless, if $P_{\mathcal{F}_0 \to \mathcal{F}_f}(t)$ exceeds 1, it is necessary to have small $\lambda$, small $p_{i,j}^{(\alpha)}$, and large $\delta E$ (if this is possible). However, owing to the analytical expression for $\Gamma_{i_0, j_0; \alpha_0}$, see Equation 9.22, it is clear that if the values of the $n_{j,\alpha}$s and of the $k_j$s are large, the approximation may likely break down very soon when $t$ increases. We have already seen something similar in the previous section when the FPL approximation broke down exactly for the same reason. This is a common feature to all the perturbation schemes that exist in the literature, both at a classical and at a quantum level: they work well (or even very well) in a certain regime (small $t$, suitable initial conditions, etc.) but when the regime changes, the approximations do not work any longer and some alternative approach should be constructed.

Let us now finally see what can be said about the portfolio of the trader $\tau_l$, which was our main interest since the beginning of Chapter 7. Assuming that we know the initial state of the system, we clearly know, in particular, the value of $\tau_l$s portfolio at time zero (actually, we know the values of the portfolios of all the traders!): extending our original definition, as we have more kinds of shares, we have $\hat{\Pi}_l(0) = \sum_{\alpha=1}^{L} P_\alpha(0) \hat{n}_{l,\alpha}(0) + \hat{k}_l(0)$. Formula (9.21) gives the transition probability from $\varphi_{\mathcal{F}_0}$ to $\varphi_{\mathcal{F}_f}$. This probability is just a single contribution in the computation of the transition probability from a given $\hat{\Pi}_l(0)$ to a certain $\hat{\Pi}_l(t)$ as the same value of the portfolio of the $l$th trader could be recovered at time $t$ for very many different states $\varphi_{\mathcal{F}_j}$: all the sets $\mathcal{G}$ with the same values of $n_{l,\alpha}^f$ and $k_l^f$, and with any other possible choice of $n_{l',\alpha}^f$ and $k_{l'}^f$, $l' \neq l$, give rise to the same value of the portfolio for $\tau_l$. Hence, if we call $\tilde{\mathcal{F}}$ the set of all these sets, we just have to sum up over all these different contributions:

$$P_{\hat{n}_l^o \to \hat{n}_l^f}(t) = \sum_{\mathcal{G} \in \tilde{\mathcal{F}}} P_{\mathcal{F}_0 \to \mathcal{G}}(t). \qquad (9.27)$$

Hence, the transition probability could be, at least formally, computed at the desired order in powers of $\lambda$.

## 9.2.2 Second-Order Corrections

We now want to show what happens going to the next order in the perturbation expansion. For that, we start considering the easiest situation, that is, the case of a time-independent perturbation $H_I$: the prices are constant in time. Hence, the integrals in formula (9.18) can be easily computed and the result is the following:

$$c_{\mathcal{F}_f}^{(2)}(t) = \sum_{\mathcal{F}} h_{\mathcal{F}_f,\mathcal{F}}(0) h_{\mathcal{F},\mathcal{F}_0}(0) \mathcal{E}_{\mathcal{F},\mathcal{F}_0,\mathcal{F}_f}(t), \qquad (9.28)$$

where

$$\mathcal{E}_{\mathcal{F},\mathcal{F}_0,\mathcal{F}_f}(t) = \frac{1}{E_{\mathcal{F}} - E_{\mathcal{F}_0}} \left( \frac{e^{i(E_{\mathcal{F}_f} - E_{\mathcal{F}_0})t} - 1}{E_{\mathcal{F}_f} - E_{\mathcal{F}_0}} - \frac{e^{i(E_{\mathcal{F}_f} - E_{\mathcal{F}})t} - 1}{E_{\mathcal{F}_f} - E_{\mathcal{F}}} \right).$$

Recalling Equation 9.19, we rewrite Equation 9.28 as

$$c_{\mathcal{F}_f}^{(2)}(t) = \sum_{\mathcal{F}} < \varphi_{\mathcal{F}_f}, H_I \varphi_{\mathcal{F}} > < \varphi_{\mathcal{F}}, H_I \varphi_{\mathcal{F}_0} > \mathcal{E}_{\mathcal{F},\mathcal{F}_0,\mathcal{F}_f}(t),$$

which explicitly shows that up to the second order in $\lambda$, transitions between states that differ for two shares are allowed—it is enough that some intermediate state $\varphi_{\mathcal{F}}$ differs for (e.g., plus) one share from $\varphi_{\mathcal{F}_0}$ and for (e.g., minus) one share from $\varphi_{\mathcal{F}_f}$.

If the $P_\alpha(t)$s depend on time, the situation is a bit more complicated but not particularly different. Going back to Example 9.2, and considering then that simple (but not completely trivial) situation in which the prices of the shares really change few times, we can perform the computation and we find

$$
\begin{aligned}
c^{(2)}_{\mathcal{F}_f}(t) = (-\mathrm{i})^2 \sum_{\mathcal{F}} \{ & h_{\mathcal{F}_f,\mathcal{F}}(t_0) h_{\mathcal{F},\mathcal{F}_0}(t_0) J_0(\mathcal{F}, \mathcal{F}_0, \mathcal{F}_f; t_1) \\
& + h_{\mathcal{F}_f,\mathcal{F}}(t_1) \left[ h_{\mathcal{F},\mathcal{F}_0}(t_0) I_0(\mathcal{F}, \mathcal{F}_0; t_1) I_1(\mathcal{F}_f, \mathcal{F}; t_2) \right. \\
& \left. + h_{\mathcal{F},\mathcal{F}_0}(t_1) J_1(\mathcal{F}, \mathcal{F}_0, \mathcal{F}_f; t_2) \right] \\
& + h_{\mathcal{F}_f,\mathcal{F}}(t_2) \left[ (h_{\mathcal{F},\mathcal{F}_0}(t_0) I_0(\mathcal{F}, \mathcal{F}_0; t_1) \right. \\
& + h_{\mathcal{F},\mathcal{F}_0}(t_1) I_1(\mathcal{F}, \mathcal{F}_0; t_2)) I_2(\mathcal{F}_f, \mathcal{F}; t_3) + \\
& \left. + h_{\mathcal{F},\mathcal{F}_0}(t_2) J_2(\mathcal{F}, \mathcal{F}_0, \mathcal{F}_f; t_3) \right] \},
\end{aligned}
\tag{9.29}
$$

where we have introduced the functions

$$
I_j(\mathcal{F}, \mathcal{G}; t) := \int_{t_j}^{t} \mathrm{e}^{\mathrm{i}(E_{\mathcal{F}} - E_{\mathcal{G}})t'} \, \mathrm{d}t' = \frac{1}{\mathrm{i}(E_{\mathcal{F}} - E_{\mathcal{G}})} \left( \mathrm{e}^{\mathrm{i}(E_{\mathcal{F}} - E_{\mathcal{G}})t} - \mathrm{e}^{\mathrm{i}(E_{\mathcal{F}} - E_{\mathcal{G}})t_j} \right)
$$

and

$$
J_j(\mathcal{F}, \mathcal{G}, \mathcal{L}; t) := \int_{t_j}^{t} I_j(\mathcal{F}, \mathcal{G}; t') \, \mathrm{e}^{\mathrm{i}(E_{\mathcal{L}} - E_{\mathcal{F}})t'} \, \mathrm{d}t'.
$$

Needless to say, this last integral could be explicitly computed, but we will not show here the explicit result because this is not very interesting for us.

The same comments as earlier about the possibility of having a non-zero transition probability can be repeated also for Equation 9.29: it is enough that the time-depending perturbation *connects* $\varphi_{\mathcal{F}_0}$ to $\varphi_{\mathcal{F}_f}$ via some intermediate state $\varphi_{\mathcal{F}}$ in a single time subinterval in order to permit a transition. If this never happens in $[0, t]$, the transition probability is zero. We can see the problem from a different point of view: if some transition takes place in the interval $[0, t]$, there must be another state, $\varphi_{\mathcal{F}'_f}$, different from $\varphi_{\mathcal{F}_f}$, such that some of the matrix elements that enter the computation of the transition probability $P_{\mathcal{F}_0 \to \mathcal{F}'_f}(t)$ are different from zero. Of course, also at this order in $\lambda$, we can recover the probability $P_{\hat{n}^o_l \to \hat{n}^f_l}(t) = \sum_{\mathcal{G} \in \tilde{\mathcal{F}}} P_{\mathcal{F}_0 \to \mathcal{G}}(t)$ as in Equation 9.27.

### 9.2.3 Feynman Graphs

Following the book of Messiah (1962), we now try to connect the analytical expression of a given approximation of $c_{\mathcal{F}_f}(t)$ with some kind of *Feynman graph* in such a way that the higher orders could be easily deduced, in principle, considering a certain set of rules, which we obviously call *Feynman rules*.

The starting point is given by the expressions in Equations 9.17 and 9.18 for $c_{\mathcal{F}_f}^{(1)}(t)$ and $c_{\mathcal{F}_f}^{(2)}(t)$, which is convenient to rewrite in the following form:

$$c_{\mathcal{F}_f}^{(1)}(t) = -\mathrm{i} \int_0^t e^{\mathrm{i}E_{\mathcal{F}_f}t_1} < \varphi_{\mathcal{F}_f}, H_I(t_1)\varphi_{\mathcal{F}_0} > e^{-\mathrm{i}E_{\mathcal{F}_0}t_1}\, dt_1, \qquad (9.30)$$

and

$$c_{\mathcal{F}_f}^{(2)}(t) = (-\mathrm{i})^2 \sum_{\mathcal{F}} \int_0^t dt_2 \int_0^{t_2} dt_1\, e^{\mathrm{i}E_{\mathcal{F}_f}t_2} < \varphi_{\mathcal{F}_f}, H_I(t_2)\varphi_{\mathcal{F}} > e^{-\mathrm{i}E_{\mathcal{F}}t_2}$$
$$\times e^{\mathrm{i}E_{\mathcal{F}}t_1} < \varphi_{\mathcal{F}}, H_I(t_1)\varphi_{\mathcal{F}_0} > e^{-\mathrm{i}E_{\mathcal{F}_0}t_1}.$$

$$(9.31)$$

The reason this is so useful is that, as we will now sketch, the different ingredients needed to find the Feynman rule are now explicitly separated and, therefore, easily identified. A graphical way to describe $c_{\mathcal{F}_f}^{(1)}(t)$ is given in Figure 9.4: at $t = t_0$, the state of the system is $\varphi_{\mathcal{F}_0}$, which evolves freely (and therefore, $e^{-\mathrm{i}E_{\mathcal{F}_0}t_1}\varphi_{\mathcal{F}_0}$ appears) until the interaction occurs, at $t = t_1$. After the interaction the system is moved to the state $\varphi_{\mathcal{F}_f}$, which evolves again freely (and therefore, $e^{-\mathrm{i}E_{\mathcal{F}_f}t_1}\varphi_{\mathcal{F}_f}$ appears, and the different sign in Equation 9.30 is due to the antilinearity of the scalar product in the first variable). The free evolutions are represented by the upward inclined arrows, while the interaction between the initial and the final states, $< \varphi_{\mathcal{F}_f}, H_I(t_1)\varphi_{\mathcal{F}_0} >$, is described by the horizontal wavy line in Figure 9.4. Obviously, as the interaction may occur at any time between 0 and $t$, we have to integrate on all these possible $t_1$s and multiply the result for $-\mathrm{i}$, which is a sort of normalization constant.

In a similar way, we can construct the Feynman graph for $c_{\mathcal{F}_f}^{(2)}(t)$, $c_{\mathcal{F}_f}^{(3)}(t)$, and so on. For example, $c_{\mathcal{F}_f}^{(2)}(t)$ can be deduced by a graph like the one in Figure 9.5, where two interactions occur, the first at $t = t_1$ and the second at $t = t_2$.

Because of the double interaction, we now have to integrate twice the result, recalling that $t_1 \in (0, t_2)$ and $t_2 \in (0, t)$. For the same reason, we have to sum over all the possible intermediate states, $\varphi_{\mathcal{F}}$. The free time evolution for the various free fields also appear in formula (9.31), as well as the normalization factor $(-\mathrm{i})^2$. Following these same rules, we could also give a formal expression for the other coefficients, $c_{\mathcal{F}_f}^{(3)}(t)$, $c_{\mathcal{F}_f}^{(4)}(t)$, and so on: the third-order correction $c_{\mathcal{F}_f}^{(3)}(t)$ contains, for instance, a double

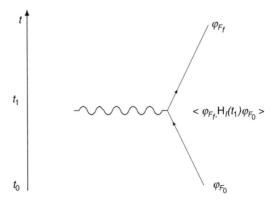

**Figure 9.4**    Graphical expression for $c^{(1)}_{\mathcal{F}_f}(t)$.

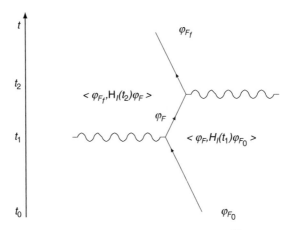

**Figure 9.5**    Graphical expression for $c^{(2)}_{\mathcal{F}_f}(t)$.

sum on the intermediate states, allowing in this way a transition from a state with, say, $n^o_{i,\alpha}$ shares to a state with $n^f_{i,\alpha} = n^o_{i,\alpha} + 3$ shares, a triple time integral, and a factor $(-i)^3$.

## 9.3    FINAL CONSIDERATIONS

It is now time to compare our approach with others proposed in the literature. Before starting, however, we want to stress that our analysis is rather *biased*: we just consider other approaches to stock markets somehow based on, or related to, quantum mechanics, discussing the main similarities and the differences with respect to our settings. The reason to limiting this analysis to these kinds of approaches is that we believe that most of the models proposed for stock markets, as well as our proposal, are still too far from *real life* to make it meaningful to compare them with real markets: nevertheless, they are, quite often, interesting, or even quite interesting, under some aspects, but they are still not able to explain in detail all the mechanisms of a market. This is, in a certain sense, useful for curious researchers because it leaves room for other ideas and different strategies.

Let us start this short review with the work of Ye and Huang (2008), in which the wave functions of a quantum harmonic oscillator are used, introducing ad hoc some decay, to discuss the *price* and the *value* of a given share. The idea is that the price of a share *fluctuates* around its value: if the price is higher than the value, most traders will try to sell shares (and this will make the price to decrease toward its value). On the other hand, if the price is smaller than the value, most traders will try to buy shares (and this will make the price to increase, again toward its value). However, the value of the share is not known with absolute precision, as otherwise no fluctuation is expected, while fluctuations are always observed in stock markets. Hence, a quantum mechanical wave function could be used to give the probability of having a certain price for the share, and to give the value of the share as well. The time evolution of the price is described in Schrödinger picture, and a damping term in the wave function is introduced by considering an *effective* nonunitary time evolution, similar to that considered in Section 2.7.1. With respect to our approach, here nothing is said about the portfolio of the traders: the interest of the authors is in the wave function describing the price and in its dynamical behavior.

The price of the share is the main object of the investigation also in Choustova (2007a, 2007b). The idea is that this price is determined by two different contributions, *hard conditions* (natural resources, industrial production, and so on) and *information factors*. The hard conditions are described by a classical potential $V(t, q_1, q_2, \ldots, q_n)$, where $q_j$ is the price of the $j$th share and $t$ is the time. The information is provided, adopting Bohm's point of view on quantum mechanics, by a pilot wave function $\Psi(t, q_1, q_2, \ldots, q_n)$, satisfying a Schrödinger equation of motion with potential $V$. More in detail, the square modulus of $\Psi$ produces, see Choustova (2007a), a second potential $U(t, q_1, q_2, \ldots, q_n)$, and the price of the $j$th share $p_j$ is found solving the following Newton-like equation: $\dot{p}_j = -\frac{\partial V}{\partial q_j} - \frac{\partial U}{\partial q_j}$. The function $g_j = -\frac{\partial U}{\partial q_j}$ is called *financial mental force*. Hence, we see that a mixture of classical and quantum mechanics is intrinsic in this analysis: classical mechanics for hard effects and quantum mechanics for information.

A similar approach, based again on Bohm's quantum mechanics, is discussed in Haven (2010) and Ishio and Haven (2009), where, as Choustova (2007a, 2007b), a pilot wave function obeying a Schrödinger equation is used to describe the information reaching the traders. An extension of the Black–Scholes equation, where this information wave function is taken into account, is proposed. Essentially in the same direction goes Khrennikov's work, see Khrennikov (2010), for instance. Incidentally, let us observe that we have also considered several times, along this book, the role of the information reaching (and being exchanged between) the actors of the system under consideration. However, in our approach, rather than being a dynamic variable, the information is usually related to the parameters appearing in the free Hamiltonian, which are fixed in time. Therefore, a possible natural extension of our settings would consist in replacing these constants with some known functions of time, such as external classical fields or, even better, with some unknown variables whose time evolution should be determined by the model itself. This particular point of view, in a certain sense, has been already considered here, and in particular in Section 8.1, where the time-dependent market supply $\hat{O}$ is directly connected with the information. Recall that the time evolution $\hat{O}(t)$ is deduced using the Heisenberg equation of motion. In other words, $\hat{O}(t)$ is, for us, the analogous of the financial mental force introduced in, for instance, Choustova (2007a, 2007b). A different possibility, only partially explored, is to associate information with reservoirs, as we have done in Part I of this book. In fact, a reservoir could also be seen as a *source of information* for the traders.

In the work by Ataullah et al. (2009), the authors use the wave function of a frictionless particle moving in a (somehow modified) one-dimensional square well, in the analysis of stock markets, working in the Schrödinger picture. They claim that the modulus square of this wave function is very similar to the instantaneous return on the *financial times stock exchange all share index*.

In the past 10 years or so, E. W. Piotrowski, J. Sładkowski, and their collaborators produced a lot of deep work on possible connections between quantum mechanics and stock markets, analyzing *quantum barguining games* in connection with quantum game theory and quantum computing. We only cite here the work of Piotrowski and Słandkowski (2005), in which the authors describe in some detail the functional structure (Hilbert space, vectors, operators) used in their analysis, which strongly reminds, mutatis mutandis, our general framework. In particular, for instance, the vectors are what the authors call *quantum strategies*, and the tactics are represented by unitary operators acting on the Hilbert space of the theory. The square moduli of the wave functions are probabilities of some action to be performed. They also use the quantum Zeno effect as a mechanism to reach an equilibrium. Such an effect could be replaced, in our approach, by the may be more realistic effect of a reservoir interacting with the system, that is, with the traders of the market. We have discussed in Part I how a reservoir can be used to model certain realistic interactions of the actors of our particular system with the external world. The same approach and a similar interpretation could be adopted for SSM.

Other than in the works by Haven (2010) and Ishio and Haven (2009), many other attempts have also been made in recent years to generalize the Black–Scholes equation to quantum settings. According to our knowledge, the first of these attempts is due to W. Segal and I. E. Segal in 1998 (Segal and Segal, 1998), in which exactly the same functional structure as the one used here has been adopted: a Hilbert space $\mathcal{H}$, a set $\mathcal{A}$ of operators acting on $\mathcal{H}$, and the states over $\mathcal{A}$. Their motivation to use operators rather than simple functions was based on the remark, already cited several times in these notes, that a given trader cannot know simultaneously the price of a share and its time derivative, as otherwise he would be able to earn a huge amount of cash. Hence the Heisenberg uncertainty principle, rather than being a problem in the description of the system, becomes the main reason why noncommuting operators should be better considered. In our opinion, rather than a *quantum approach*, what is proposed in Segal and Segal (1998) is an *operatorial point of view*. From this aspect, this is very close to what we have in mind in our approach: *quantum* is not really the important word for us, but *operator*. Other possible extensions of the Black–Scholes equation to the quantum regime can also be found in Accardi and Boukas (2006), in the context of the Hudson–Parthasarathy quantum stochastic calculus, and in Melnyk and Tuluzov (2008).

A comprehensive work on relations between quantum mechanics and economics is provided by Baaquie (2004). A theory of stock options and a modeling of interest rates are proposed using quantum mechanical tools, and in particular, Hamiltonian operators and path integrals. Not surprisingly, the Black–Scholes equation is considered within this framework, together with many other arguments such as the Heisenberg uncertainty principle and its appearance in finance. Other than this, the role of quantum field theory is discussed in connection with interest rates models.

Very close to our work is also the paper by Schaden (2002). In particular, we share with Schaden's work the framework made of a certain Hilbert space $\mathcal{H}$ and of operators acting on $\mathcal{H}$, which can be seen as selling and buying operators, a vacuum of $\mathcal{H}$ on which the operators act to produce other vectors of the same Hilbert space, and so on. The *state of the market* is a vector in $\mathcal{H}$, essentially as in our approach, which can be *pure* (it is just one of our number vectors (Eq. 2.4)), or not (it is a linear combination of number vectors), and the time evolution of the market is driven by a self-adjoint operator, the Hamiltonian of the market, which gives rise to a unitary evolution. The CCR, in a slightly modified form, also play a role similar to those played in our settings, but translation operators also appear that efficiently describe the change of cash of the various traders. Since 2002, Schaden has no longer contributed to quantum finance. This is, in our opinion, a great loss and we hope to see some new papers by him on this argument soon.

Yukalov, Sornette, and their collaborators have produced, along the years, several interesting papers on what they call the *quantum decision theory*, also in connection with stock markets (2009a,b, 2012). Once again, we have a complete coincidence of our frameworks, and this is not surprising. What is more interesting is, probably, their claim, identical to ours, that what is more important for them is the framework of quantum mechanics rather than other aspects.

We end this brief review citing a paper by Aerts et al. (2010), which describes how *quantum corrections* are needed to explain the behavior of stock markets if economical crises are to be described: using random walks may only work for ideal situations, for *efficient markets*.

The analysis given earlier, which surely excludes many other contributions hidden in the huge scientific literature, gives the feeling that there is a very strong interest in possible relations between quantum mechanics and economics and that there is still a lot of work to do to produce a reasonable *theory of stock markets,* at both the classical and quantum levels. It is, indeed, unclear to us if such a theory can even be constructed or if we can simply produce some simplified models. This is exactly what is available up to now in the different papers quickly reviewed here: they all produce nice, but partial, results. Hence, this is not the end of the story!

# CHAPTER 10

# CONCLUSIONS

In these notes, we have considered applications of our framework to many, and very different, classical systems: love affairs, competition between species, levels of welfare for bacteria, and stock markets. Many questions can be quite naturally considered. For instance, can our approach be useful in the description of other dynamical systems? Which kind of classical systems can be naturally analyzed using our operatorial approach? Are the bosonic and fermionic number operators the only possible choices? Are the rules listed in Chapter 6 the end of the story? In this chapter, we want to answer some of these questions, and in particular, we want to show how the algebraic setting used all along this book, and which is represented by CCR and CAR, can be extended and why this extension could be interesting in concrete applications.

## 10.1  OTHER POSSIBLE NUMBER OPERATORS

The main ingredient all along this book is surely the triple $(a, a^\dagger, N := a^\dagger a)$ of bosonic or fermionic annihilation, creation and number operators, obeying simple commutation rules. These are not the only possible choices, but they are surely the most natural ones. For the sake of completeness, here also, we briefly discuss other old and new possibilities,

*Quantum Dynamics for Classical Systems: With Applications of the Number Operator*,
First Edition. Fabio Bagarello.
© 2013 John Wiley & Sons, Inc. Published 2013 by John Wiley & Sons, Inc.

starting from the well-known Pauli matrices, and then introducing pseudobosons and nonlinear pseudobosons. We also briefly discuss why using these other possibilities could be interesting for us.

### 10.1.1 Pauli Matrices

The Pauli matrices produce a simple representation of the CAR. More in detail, let us consider

$$\sigma_1 = \begin{pmatrix} 0 & 1 \\ 1 & 0 \end{pmatrix}, \quad \sigma_2 = \begin{pmatrix} 0 & -i \\ i & 0 \end{pmatrix}, \quad \sigma_3 = \begin{pmatrix} 1 & 0 \\ 0 & -1 \end{pmatrix} \quad (10.1)$$

They satisfy, among others, the following anticommutation rule: $\sigma_i \sigma_j + \sigma_j \sigma_i = 2\delta_{ij} \mathbb{1}_2$, where $\mathbb{1}_2$ is the two–two identity matrix. With the above definition, it is clear that the eigenstates of $\sigma_3$ are

$$\varphi_+ = \begin{pmatrix} 1 \\ 0 \end{pmatrix}, \quad \text{and} \quad \varphi_- = \begin{pmatrix} 0 \\ 1 \end{pmatrix},$$

with eigenvalues $+1$ and $-1$, respectively: $\sigma_3 \varphi_\pm = \pm\varphi_\pm$. It is clear that defining $\hat{n} := \frac{1}{2}(\mathbb{1}_2 - \sigma_3) = \begin{pmatrix} 0 & 0 \\ 0 & 1 \end{pmatrix}$, this behaves exactly as $a^\dagger a$ for fermionic $a$: $\{a, a^\dagger\} = \mathbb{1}_2$. This, in particular, implies that $\varphi_+$ could be considered as the vacuum of $a$, while $\varphi_-$ is the (only) excited state. Moreover, it is a simple exercise to relate $a$ and $a^\dagger$ with the Pauli matrices $\sigma_j$:

$$\sigma_1 = a + a^\dagger, \quad \sigma_2 = i(a^\dagger - a), \quad \sigma_3 = \mathbb{1}_2 - 2a^\dagger a.$$

This representation of the CAR has a consequence that is interesting for us: the model described in Chapter 4 could be rewritten entirely in terms of spin operators. Hence, at least at a formal level, we could introduce a Hamiltonian operator in terms of Pauli matrices. As discussed, for instance, in Stanley et al. (2001), spin operators are interesting also in connection with economics, even because they produce many-body systems for which it is reasonably easy, or at least possible and reasonable, to carry on some statistical analysis, considering, for instance, their phase transitions. This is an exciting possibility for the future: understand our model in terms of spin variables, and consider it from a statistical point of view, trying to see if such an approach enables us or not to obtain new interesting aspect of the system.

### 10.1.2 Pseudobosons

Another possible way to get other number operators is provided by the so-called pseudobosons; see Bagarello (2011a). They arise from an extension of the CCR, which is replaced by the commutation rule $[A, B] = \mathbb{1}$, with $B \neq A^\dagger$. We have shown that under suitable conditions that are quite often satisfied in concrete systems, $A$, $B$, $N := BA$, and $N^\dagger = A^\dagger B^\dagger$ produce two biorthogonal families of vectors in $\mathcal{H}$, $\mathcal{F}_\varphi := \{\varphi_n, n \geq 0\}$ and $\mathcal{F}_\Psi := \{\Psi_n, n \geq 0\}$, which generate all of $\mathcal{H}$. These vectors are respectively eigenstates of $N$ and $N^\dagger$ with eigenvalue $n$: $N\varphi_n = n\varphi_n$ and $N^\dagger \Psi_n = n\Psi_n$, for all $n \geq 0$. The framework extends that originated by CCR, to which it reduces when $B$ and $A^\dagger$ coincide. In this case, in particular, it turns out that $N = N^\dagger$ and that $\mathcal{F}_\varphi \equiv \mathcal{F}_\Psi$. This approach may play some relevant role in those situations in which the dynamics of the system is described by a non-self-adjoint Hamiltonian, as, for instance, proposed in Romero et al. (2011).[1] Furthermore, the existence of two non-self-adjoint operators suggests the possibility of using pseudobosons to construct some effective Hamiltonian useful to describe a nonunitary time evolution as the one briefly considered in Section 2.7. This might be relevant to avoid the use of the otherwise computational heavy (but physically motivated) mechanism of the reservoirs used before to get some decay in the dynamics of observables quantities. This could be particularly interesting in the analysis of SSM, where the system is already very complicated even with no reservoir at all.

### 10.1.3 Nonlinear Pseudobosons

An extended version of pseudobosons, giving rise to *generalized* number operators, can be introduced starting from a strictly increasing sequence $\{\epsilon_n\}$ such that $\epsilon_0 = 0$: $0 = \epsilon_0 < \epsilon_1 < \cdots < \epsilon_n < \cdots$. Then, given two operators $a$ and $b$ on the Hilbert space $\mathcal{H}$, in Bagarello (2011b), we have introduced the following.

**Definition 10.1**  *We will say that the triple* $(a, b, \{\epsilon_n\})$ *is a family of nonlinear regular pseudobosons (NLRPB) if the following properties hold:*

- **p1.** *a nonzero vector* $\Phi_0$ *exists in* $\mathcal{H}$ *such that* $a\,\Phi_0 = 0$ *and* $\Phi_0 \in D^\infty(b)$.[2]
- **p2.** *a nonzero vector* $\eta_0$ *exists in* $\mathcal{H}$ *such that* $b^\dagger \eta_0 = 0$ *and* $\eta_0 \in D^\infty(a^\dagger)$.

---

[1] The idea is to replace the self-adjoint Hamiltonian of, say, a quantum harmonic oscillator, $H = a^\dagger a + \frac{1}{2}\mathbb{1}$, with its pseudobosonic extension, $h = BA + \frac{1}{2}\mathbb{1}$. It is clear that $h \neq h^\dagger$, so that the related dynamics is not unitary and it could describe damping.

[2] Here $D^\infty(b)$ is the domain of all the powers of the operator $b$.

- **p3.** *Calling*

$$\Phi_n := \frac{1}{\sqrt{\epsilon_n!}} b^n \Phi_0, \qquad \eta_n := \frac{1}{\sqrt{\epsilon_n!}} a^{\dagger n} \eta_0, \qquad (10.2)$$

*we have, for all* $n \geq 0$,

$$a \, \Phi_n = \sqrt{\epsilon_n} \, \Phi_{n-1}, \qquad b^\dagger \eta_n = \sqrt{\epsilon_n} \, \eta_{n-1}. \qquad (10.3)$$

- **p4.** *The sets* $\mathcal{F}_\Phi = \{\Phi_n, \, n \geq 0\}$ *and* $\mathcal{F}_\eta = \{\eta_n, \, n \geq 0\}$ *are bases of* $\mathcal{H}$.
- **p5.** $\mathcal{F}_\Phi$ *and* $\mathcal{F}_\eta$ *are Riesz bases of* $\mathcal{H}$.[3]

In particular, if $\epsilon_n = n$, we recover ordinary pseudobosons: $[a, b] = 1$. Let us introduce the following (not self-adjoint) operators:

$$M = ba, \qquad \mathcal{M} = M^\dagger = a^\dagger b^\dagger. \qquad (10.4)$$

Then, we can check that $\Phi_n \in D(M) \cap D(b)$, $\eta_n \in D(\mathcal{M}) \cap D(a^\dagger)$, and, more than this, that

$$b \, \Phi_n = \sqrt{\epsilon_{n+1}} \, \Phi_{n+1}, \qquad a^\dagger \eta_n = \sqrt{\epsilon_{n+1}} \, \eta_{n+1}, \qquad (10.5)$$

which is a consequence of definitions (Eq. 10.2), as well as

$$M \, \Phi_n = \epsilon_n \Phi_n, \qquad \mathcal{M} \, \eta_n = \epsilon_n \eta_n, \qquad (10.6)$$

These eigenvalue equations have a very important consequence: the vectors in $\mathcal{F}_\Phi$ and $\mathcal{F}_\eta$ are mutually orthogonal. More explicitly,

$$\langle \Phi_n, \eta_m \rangle = \delta_{n,m} \langle \Phi_0, \eta_0 \rangle. \qquad (10.7)$$

Continuing our analysis on the consequences of the definition of NLRPB, and in particular of **p4**, we rewrite this assumption in bra-ket formalism as

$$\sum_n |\Phi_n> < \eta_n| = \sum_n |\eta_n> < \Phi_n| = \mathbb{1}, \qquad (10.8)$$

whereas **p5** implies that if $\langle \Phi_0, \eta_0 \rangle = 1$, the operators $S_\Phi := \sum_n |\Phi_n> < \Phi_n|$ and $S_\eta := \sum_n |\eta_n> < \eta_n|$ are positive, bounded, and invertible.

---

[3] Recall that a set of vectors $\phi_1, \phi_2, \phi_3, \ldots$, is a Riesz basis of a Hilbert space $\mathcal{H}$, if there exists a bounded operator $V$, with bounded inverse and an o.n. basis of $\mathcal{H}$, $\varphi_1, \varphi_2, \varphi_3, \ldots$, such that $\phi_j = V\varphi_j$, for all $j = 1, 2, 3, \ldots$

Moreover, it is possible to show that $S_\Phi = S_\eta^{-1}$. The new fact is that the operators $a$ and $b$ do not, in general, satisfy any *simple* commutation rule. Indeed, we can check that for all $n \geq 0$,

$$[a, b]\Phi_n = \left(\epsilon_{n+1} - \epsilon_n\right)\Phi_n, \qquad (10.9)$$

which returns $[a, b]\Phi_n = \Phi_n$, if $\epsilon_n = n + \alpha$, for some real $\alpha$, but not in general.

More details can be found in Bagarello (2011b). Here, we are more interested in the possible applications of NLRPB within our context. The idea is very simple: rather than the CCR, we could use NLRPB in the nonlinear models discussed in these notes. For instance, in Part II, we have always worked under the simplifying assumption that the traders can only exchange one share in each transaction. This is because the eigenvalues of $N = a^\dagger a$ are just the natural numbers, if $a$ satisfies the CCR. If we replace $N$ with, say, the operator $M$ in Equation 10.4, using $a$ and $b$ satisfying Definition 10.1, then, because of Equation 10.6, we could model a situation in which not just one, but $\epsilon_n$ shares are exchanged between the traders in a single transaction. Analogously, we could use these operators also in the nonlinear models considered in Chapter 3, to replace the operators $a_1^M$ and its adjoint appearing, for instance, in Definition 3.5, with some nonlinear pseudobosonic operator. The main problem, however, is in the commutation rules involving $a$, $b$, $M$, and $M^\dagger$, which could be very difficult, making this strategy not easy to be implemented in concrete models, but surely interesting to be considered.

Other operators have been proposed to model, for instance, the fact that the money of a trader can increase or decrease by several units of cash (Schaden, 2002). Schaden uses a unitary displacement operator, very close to the translation operator $e^{i\alpha p}$, $\alpha \in \mathbb{R}$, $[x, p] = i\,\mathbb{1}$, acting on the wave function of the system as follows: $e^{i\alpha p} f(x) = f(x + \alpha)$, for each square integrable $f(x)$. This is a slight variation on the same theme of CCR as $p$ can be written in terms of bosonic operators as $p = \frac{a - a^\dagger}{\sqrt{2}\,i}$. In our opinion, this is an interesting suggestion that deserves a deeper analysis, because it surely simplifies the operators describing the system, even if, at the same time, makes more complicated the construction of the vector of the system, as it becomes a tensor product of a wave function and of a number vector.

### 10.1.4  Algebra for an $M + 1$ Level System

In Chapter 3, to get a finite dimensional Hilbert space with dimension different from 2 (the fermionic case), we have used CCR and then we have *truncated* the Hilbert space, going from the infinite dimensional $\mathcal{H}$ to a finite dimensional version, $\mathcal{H}_{\text{eff}}$. The reason for keeping the bosonic operators was mainly that they obey easy commutation rules, so that the deduction of the equations of motion was nothing but a simple exercise. However, introducing such a cutoff in the dimension of the Hilbert space is not always an easy task as it may give rise to extra difficulties. For this reason, it might be useful to look for some operators that act as raising and lowering operators in a Hilbert space, which is genuinely of dimension $M + 1$, with $2 \leq M < \infty$. This is possible.

***Using an o.n. Basis as Starting Point***   Let $\mathcal{H}$ be an $M + 1$ dimensional Hilbert space, and $\mathcal{F}_\varphi := \{\varphi_0, \varphi_1, \ldots, \varphi_M\}$ be an o.n. basis for $\mathcal{H}$. Let us define the operators

$$a_\uparrow := \sum_{k=0}^{M-1} |\varphi_{k+1}\rangle \langle \varphi_k|,$$

and its adjoint

$$a_\uparrow^\dagger = \sum_{k=0}^{M-1} |\varphi_k\rangle \langle \varphi_{k+1}| =: a_\downarrow.$$

We refer to Section 3.2.1 for the meaning of these operators, which are just sums of projectors. They satisfy the following equations:

$$a_\uparrow \varphi_l = \begin{cases} \varphi_{l+1}, & l = 0, 1, \ldots, M - 1 \\ 0 & l = M, \end{cases}$$

and

$$a_\downarrow \varphi_l = \begin{cases} \varphi_{l-1}, & l = 1, 2, \ldots, M \\ 0 & l = 0. \end{cases}$$

It is natural to interpret these operators respectively as rising and lowering operators for our finite dimensional Hilbert space $\mathcal{H}$. The identity operator on $\mathcal{H}$ can be written, adopting again the Dirac notation, as $\mathbb{1} = \sum_{k=0}^M |\varphi_k\rangle\langle\varphi_k|$, while the number operator is

$$\hat{N} = \sum_{k=0}^M k\, |\varphi_k\rangle \langle \varphi_k|,$$

which satisfies the eigenvalue equation $\hat{N} \varphi_l = l \varphi_l$. Therefore, everything works perfectly, up to this point. However, when we compute the commutation rule between $a_\uparrow$ and $a_\downarrow$, the problem arises. Indeed, we find that

$$\left[a_\uparrow, a_\downarrow\right] = |\varphi_M\rangle \langle\varphi_M| - |\varphi_0\rangle \langle\varphi_0|,$$

which is a difference of orthogonal projection operators on the two *extreme* levels of the Hilbert space.

***Using CAR as Starting Point*** If the dimensionality of the Hilbert space we are interested in is $2^K$, for some $K = 2, 3, \ldots$, we can use CAR for constructing the rising, lowering, and number operators. For instance, suppose that we are interested in a four-dimensional Hilbert space. Then, we consider two independent fermionic modes, whose corresponding operators $b_1$ and $b_2$ satisfy the CAR:

$$\{b_i, b_j^\dagger\} = \delta_{i,j}\, \mathbb{1}, \qquad b_i^2 = 0,$$

$i = 1, 2$. Let $\varphi_{0,0}$ be the vacuum of these operators: $b_1 \varphi_{0,0} = b_2 \varphi_{0,0} = 0$. We can introduce the following vectors, which form an o.n. basis of $\mathcal{H}$:

$$\Phi_0 := \varphi_{0,0}, \quad \Phi_1 := b_1^\dagger \varphi_{,0}, \quad \Phi_2 := b_2^\dagger \varphi_{0,0}, \quad \Phi_3 := b_1^\dagger b_2^\dagger \varphi_{0,0}.$$

The number operator associated with these vectors is easily found: $\hat{N} := b_1^\dagger b_1 + 2\, b_2^\dagger b_2$. Indeed we have $\hat{N}\, \Phi_l = l\, \Phi_l$. Again, what become more complicated are the anticommutation rules because the raising and lowering operators are not the ones as in a two-dimensional (single fermionic mode) Hilbert space. However, as these rules can be deduced by those for a single mode, this approach seems to be *more friendly* with respect to the previous one.

These are just some possibilities for extending our settings. It seems to us that each possible approach gives rise to several new problems to be solved and questions to be answered. Some interesting physics can be deduced and some interesting mathematics might be involved: more work to do!

## 10.2 WHAT ELSE?

All along these notes, we have discussed many things that are still to be done to make our models more realistic. Let us recall some: we have understood that the parameters appearing in most free Hamiltonians proposed here are related to the amount of information reaching the actors of the game. We have considered these parameters to be constant in time. A natural question is: what if these parameters are replaced by time-dependent functions? Or, in other words, can the information be included in the models as a dynamic variable? This was partly done in Chapter 8, using the market supply, but the general strategy is still to be fully understood. A second interesting extension, much simpler, concerns the different topologies in Chapter 4, representing different biological environments. This analysis will be based simply on different choices of the diffusion coefficients $p_{\alpha,\beta}$, and will be considered soon. Also, why not considering some time dependence in the $p_{\alpha,\beta}$s? But, going back to harder problems: does our strategy suggest some simple mechanism to fix the price of the shares in a *slightly more realistic* stock market? Or, stated in different words, is there any reasonable way to produce a dynamics for the market supply operator? Another major aspect to be still considered is the role of the uncertainty relation in concrete applications. As we have already seen, this is sometimes considered as the main reason to use quantum ideas in classical contexts; see Segal and Segal (1998). Surely, this will be part of our future interests.

These are just some of the possible points to be considered. Another question is the following: is it possible to use CCR, CAR, or their extensions, in the analysis of other classical systems? In our opinion, the answer is definitely yes. Just to cite another, quite trendy, application, which we have not considered yet, we believe that some social networks can be described in terms of fermionic, or spin, operators. It is also not difficult to imagine a possible application of our framework in other biological systems (e.g., analysis of fish stocks). On a different side, it could also be quite interesting, as discussed here, to use the Hamiltonians introduced all along these notes, or at least some of them, to analyze the statistical properties of the underlying systems. These would produce a completely new subject, not yet considered, which could produce interesting results and could open new lines of research. Also, we do not exclude that different interpretations of the differential equations arising from a given Hamiltonian could also exist, the interpretation of which should simplify some of the technical difficulties described in these notes. This is part of our work in progress.

So, this is just the end of *part one* of this movie. We hope *part two* will begin soon!

# BIBLIOGRAPHY

D. Abbott, P. C. W. Davies, A. K. Pati, *Quantum Aspects of Life*, Imperial College Press, London, 2008.

L. Accardi, A. Boukas, *The quantum black-scholes equation*, Glob. J. Pure Appl. Math., **2**(2), 155–170, 2006.

L. Accardi, Y. G. Lu, I. Volovich, *Quantum Theory and its Stochastic Limit*, Springer, Berlin, 2002.

D. Aerts, *A potentiality and conceptuality interpretation of quantum mechanics*, Philosophica, **83**, 15–52, 2010.

D. Aerts, S. Aerts, L. Gabora, *Experimental Evidence for Quantum Structure in Cognition*, Lecture Notes in Artificial Intelligence; 5494, Proceedings of the 3rd International Symposium on Quantum Interaction, 59–70, 2009.

D. Aerts, B. D'Hooghe, S. Sozzo, *A quantum-like approach to the stock market*, AIP Conf. Proc., 1424, 495–506, 2012.

D. Aerts, B. D'Hooghe, E. Haven, *Quantum experimental data in psychology and economics*, Int. J. Theor. Phys., **49**, 2971–2990, 2010.

V. Alfi, A. De Martino, L. Pietronero, A. Tedeschi, *Detecting the traders' strategies in minority-majority games and real stock-prices*, Physica A, **382**, 1, 2007.

M. Arndt, T. Juffmann, V. Vedral, *Quantum physics meets biology*, HFSP J., **3**(6), 386400, 2009.

*Quantum Dynamics for Classical Systems: With Applications of the Number Operator*,
First Edition. Fabio Bagarello.
© 2013 John Wiley & Sons, Inc. Published 2013 by John Wiley & Sons, Inc.

A. Ataullah, I. Anderson, M. Tippett, *A wave function for stock market returns*, Physica A, **388**, 455–461, 2009.

B. E. Baaquie, *Quantum Finance*, Cambridge University Press, Cambridge, England, 2004.

F. Bagarello, *Many-body applications of the stochastic limit: a review*, Rep. Math. Phys., **56**(1), 117–152, 2005.

F. Bagarello, *An operatorial approach to stock markets*, J. Phys. A, **39**, 6823–6840, 2006.

F. Bagarello, *Stock markets and quantum dynamics: a second quantized description*, Physica A, **386**, 283–302, 2007.

F. Bagarello, *Multiplication of distributions in any dimension: applications to δ-function and its derivatives*, J. Math. Anal. Appl., **337**(2), 1337–1344, 2008.

F. Bagarello. *Simplified stock markets described by number operators*, Rep. Math. Phys., **63**(3), 381–398, 2009a.

F. Bagarello, *A quantum statistical approach to simplified stock markets*, Physica A, **388**, 4397–4406, 2009b.

F. Bagarello, *Pseudo-bosons, so far*, Rep. Math. Phys., **68**(2), 175–210, 2011a.

F. Bagarello, *Non linear pseudo-bosons*, J. Math. Phys., **52**, 063521, 2011b.

F. Bagarello, *Damping in Quantum Love Affairs*, Physica A, **390**, 2803–2811, 2011c.

F. Bagarello, *Few simple rules to fix the dynamics of classical systems using operators*, Int. J. Theor. Phys., 2012. DOI: 10.1007/s10773-012-1085-y.

F. Bagarello, G. Morchio, *Dynamics of mean field spin models from basic results in abstract differential equations*, J. Stat. Phys., **66**, 849–866, 1992.

F. Bagarello, F. Oliveri, *Quantum Modeling of Love Affairs*. Proceedings Wascom 2009, A. M. Greco, S. Rionero, T. Ruggeri, eds, 7–14, World Scientific, Singapore, 2010.

F. Bagarello, F. Oliveri, *An operator–like description of love affairs*, SIAM J. Appl. Math., **70**, 3235–3251, 2011.

P. Ball, *The dawn of quantum biology*, Nature, **474**, 272–274, 2011.

S. M. Barnett, P. M. Radmore, *Methods in Theoretical Quantum Optics*, Clarendon Press, Oxford, 1997.

Y. Ben-Aryeh, A. Mann, I. Yaakov, *Rabi oscillations in a two-level atomic system with a pseudo-hermitian hamiltonian*, J. Phys. A, **37**, 12059–12066, 2004.

G. E. Bijwaard, Modeling migration dynamics of immigrants: the case of The Netherlands, TI 2008-070/4 Tinbergen Institute Discussion Paper, 2008.

O. Bratteli, D. W. Robinson, *Operator Algebras and Quantum Statistical Mechanics 1*, Springer-Verlag, New York, 1987a.

O. Bratteli, D. W. Robinson, *Operator Algebras and Quantum Statistical Mechanics 2*, Springer-Verlag, New York, 1987b.

E. Buffet, P. A. Martin, *Dynamics of the open BCS model*, J. Stat. Phys., **18**(6), 585–632, (1978)

J. R. Busemeyer, Z. Wang, J. T. Townsend, *Quantum dynamics of human decision-making*, J. Math. Psychol., **50**(3), 220–241, 2006.

O. Cherbal, M. Drir, M. Maamache, D. A. Trifonov, *Fermionic coherent states for pseudo-Hermitian two-level systems*, J. Phys. A, **40**, 1835–1844, 2007.

O. Choustova, *Quantum Bohmian model for financial market*, Physica A, **374**, 304–314, 2007.

O. Choustova, *Toward quantum-like modeling of financial processes*, J. Phys. A: Conf. Ser., **70**, 012006, 2007.

H. N. Comins, M. P. Hassel, *The dynamics of predation and competition in patchy environment*, Theor. Pop. Biol., **31**, 393–421, 1987.

H. N. Comins, M. P. Hassel, R. M. May, *The spatial dynamics of host–parasitoid systems*, J. Animal Ecol., **61**, 735–748, 1992.

G. S. Engel, T. S. Calhoun, E. L. Read, T.-K. Ahn, T. Mancal, Y.-C. Cheng, R. E. Blankenship, G. R. Fleming, *Evidence for wavelike energy transfer through quantum coherence in photosynthetic systems*, Nature, **446**, 782–786, 2007.

E. M. Gauger, E. Rieper, J. J. L. Morton, S. C. Benjamin, V. Vedral, *Sustained quantum coherence and entanglement in the avian compass*, Phys. Rev. Lett., **106**(4), 040503, 2011.

J. M. Gottman, J. D. Murray, C. C. Swanson, R. Tyson, K. R. Swanson, *The Mathematics of Marriage, Dynamics Nonlinear Models*, The MIT Press, Cambridge, Massachusetts, 2002.

I. Hanski, *Coexistence of competitors in patchy environment with and whitout predation*, Oikos, **37**, 306–312, 1981.

I. Hanski, *Coexistence of competitors in patchy environment*, Ecology, **64**, 493–500, 1983.

I. Hanski, *Spatial patterns of coexistence of competing species in patchy habitat*, Theor. Ecol., **1**, 29–43, 2008.

I. Hanski, M. Gilpin, *Metapopulation dynamics: brief history and conceptual domain*, Biol. J. Linn. Soc., **42**, 3–16, 1991.

E. Haven, *Pilot-wave theory and financial option pricing*, Int. J. Theor. Phys., **44**(11), 1957–1962, 2010.

H. Ishio, E. Haven, *Information in asset pricing a wave function approach*, Ann. Phys., **18** (1), 33–44, 2009.

A. R. Ives, R. M. May, *Competition within and between species in a patchy environment: relations between microscopic and macroscopic models*, J. Theor. Biol., **115**, 65–92, 1985.

E. Jimenez, D. Moya, *Econophysics: from game theory and information theory to quantum mechanics*, Physica A, **348**, 505–543, 2005.

A. Khrennikov, *Quantum-like brain: "interference of minds"*, BioSystems, **84**, 225–241, 2006.

A. Khrennikov, *Ubiquitous Quantum Structure: From Psychology to Finances*, Springer, Berlin, 2010.

T. Lux, M. Marchesi, *Scaling and criticality in a stochastic multi-agent model of a financial market*, Nature, **397**(498), 1999.

M. Makowski, E. W. Piotrowski, *Decisions in elections—transitive or intransitive quantum preferences*, J. Phys. A: Math. Theor., **44**, 215303, 2011.

R. M. Mantegna, E. Stanley, *Introduction to Econophysics*, Cambridge University Press, Cambridge, England, 1999.

P. A. Martin, *Modéles en Mécanique Statistique des Processus Irréversibles*, Lecture Notes in Physics, 103, Springer-Verlag, Berlin, 1979.

M. A. Martin-Delgado, *On quantum effects in a theory of biological evolution*, Sci. Rep. 2, 302, DOI: 10.1038/srep00302, 2012.

S. I. Melnyk, I. G. Tuluzov, *Quantum analog of the Black-Scholes formula (market of financial derivatives as a continuous weak measurement)*, Elect. J. Theor. Phys., **5**(18), 95–108, 2008.

M. B. Mensky, *Consciousness and Quantum Mechanics: Life in Parallel Worlds, Miracles of Consciousness from Quantum Reality*, World Scientific, Singapore, 2010.

E. Merzbacher, *Quantum Mechanics*, Wiley, New York, 1970.

A. Messiah, *Quantum Mechanics*, vol. 2, North Holland Publishing Company, Amsterdam, 1962.

A. Mostafazadeh, *Pseudo-hermitian representation of quantum mechanics*, Int. J. Geom. Methods Mod. Phys., **7**, 1191–1306, 2010.

J. D. Murray, *Mathematical Biology II: Spatial Models and Biomedical Applications*, Springer, Berlin, 2003.

S. Nee, R. M. Mat, *Dynamics of metapopulations: habitat destruction and competitive coexistence*, J. Animal Ecol., **61**, 37–40, 1992.

F. Oliveri, F. Paparella, *A Particle–mesh Numerical Method for Advection–reaction–diffusion Equations with Applications to Plankton Modelling*. Proceedings Wascom 2007, N. Manganaro, R. Monaco, S. Rionero, eds, World Scientific Publishing, Singapore, 469–474, 2008.

G. Panitchayangkoon, D. V. Voronine, D. Abramavicius, J. R. Caram, N. H. C. Lewis, S. Mukamel, G. S. Engel, *Direct evidence of quantum transport in photosynthetic light-harvesting complexes*, Proc. Natl. Acad. Sci. U.S.A., **108**(52), 20908–20912, 2011.

P. Pedram, *The minimal length uncertainty and the quantum model for the stock market*, Physica A, **391**(5), 2100–2105, 2012.

E. W. Piotrowski, J. Słandkowski, *Quantum diffusion of prices and profits*, Physica A, **345**, 185–195, 2005.

O. Pusuluk, C. Deliduman, *Entanglement swapping model of DNA replication*, 2011. arXiv: 1101.0073 [quant-ph].

T. Quint, M. Shubik, *A model of migration*, Cowles Foundation Discussion Paper No. 1088, 1994.

M. Reed, B. Simon, *Methods of Modern Mathematical Physics, I*, Academic Press, New York, 1980.

S. Rinaldi, *Laura and Petrarch: An Intriguing Case of Cyclical Love Dynamics*, SIAM J. Appl. Math., **58**, 1205–1221, 1998.

S. Rinaldi, *Love dynamics: the case of linear couples*, Appl. Math. Comput., **95**, 181–192, 1998.

T. Ritz, P. Thalau, J. B. Phillips, R. Wiltschko, W. Wiltschko, *Resonance effects indicate a radical-pair mechanism for avian magnetic compass*, Nature, **429**, 177–180, 2004.

P. Roman, *Advanced Quantum Mechanics*, Addison-Wesley, New York, 1965.

J. M. Romero, O. Gonzalez-Gaxiola, J. Ruiz de Chavez, R. Bernal-Jaquez, *The Black-Scholes equation and certain quantum hamiltonians*, Int. J. Pure Appl. Math., **67**(2), 165–173, 2011.

M. Schaden, *Quantum finance*, Physica A, **316**, 511–538, 2002.

W. Segal, I. E. Segal, *The Black–Scholes pricing formula in the quantum context*, Proc. Natl. Acad. Sci. U.S.A., **95**, 4072–4075, 1998.

G. L. Sewell, *Quantum Theory of Collective Phenomena*, Oxford University Press, Oxford, 1989.

G. L. Sewell, *Quantum Mechanics and its Emergent Macrophysics*, Princeton University Press, Princeton, NJ, 2002.

L. Shi, *Does security transaction volume-price behavior resemble a probability wave?* Physica A, **366**, 419–436, 2005.

L. F. Shampine, M. K. Gordon, W. H. Freeman. *Computer solution of ordinary differential equations: the initial value problem*, San Francisco, 1975.

M. Slatkin, *Competition and regional coexistence*, Ecology, **55**, 128–134, 1974.

J. C. Sproot, *Dynamical models of love*, Nonlin. Dyn. Psycol. Life Sci., **8**, 303–314, 2004.

J. C. Sproot, *Dynamical models of happiness*, Nonlin. Dyn. Psycol. Life Sci., **9**, 23–36, 2005.

H. E. Stanley, L. A. N. Amaral, X. Gabaix, P. Gopikrishnan, V. Plerou, *Similarities and differences between physics and economics*, Physica A **299**, 1–15, 2001.

R. F. Streiter, A. S. Wightman, *PCT, Spin and Statistics, and All That*, Benjamin, New York, 1964.

S. H. Strogatz, *Love affairs and differential equations*, Math. Mag., **61**, 35, 1988.

S. H. Strogatz, *Nonlinear Dynamics and Chaos*, Addison-Wesley, Reading, MA, 1994.

W. Thirring and A. Wehrl, *On the mathematical structure of the B.C.S.-Model*, Commun. Math. Phys. **4**, 303–314, 1967.

C. Ye, J. P. Huang, *Non-classical oscillator model for persistent fluctuations in stock markets*, Physica A, **387**, 1255–1263, 2008.

V. I. Yukalov, D. Sornette, *Physics of risk and uncertainty in quantum decision making*, Eur. Phys. J. B, **71**, 533–548, 2009a.

V. I. Yukalov, D. Sornette, *Processing information in quantum decision theory*, Entropy, **11**, 1073–1120, 2009b.

V. I. Yukalov, D. Sornette, *Quantum decision making by social agents*, 2012. arXiv:1202.4918v1 [physics.soc-ph].

C. Zhang, L. Huang, *A quantum model for the stock market*, Physica A, **389**(24), 5769–5775, 2010.

# INDEX

*Quantum Dynamics for Classical Systems: With Applications of the Number Operator,*
First Edition. Fabio Bagarello.
© 2013 John Wiley & Sons, Inc. Published 2013 by John Wiley & Sons, Inc.